T0269074

THE ANALYTIC S-MATRIX

THE ANALYTIC
S-MATRIX

BY

R.J.EDEN   P.V.LANDSHOFF   D.I.OLIVE

J.C.POLKINGHORNE

CAMBRIDGE
AT THE UNIVERSITY PRESS
1966

PUBLISHED BY THE PRESS SYNDICATE OF THE UNIVERSITY OF CAMBRIDGE
The Pitt Building, Trumpington Street, Cambridge, United Kingdom

CAMBRIDGE UNIVERSITY PRESS
The Edinburgh Building, Cambridge CB2 2RU, UK
40 West 20th Street, New York NY 10011–4211, USA
477 Williamstown Road, Port Melbourne, VIC 3207, Australia
Ruiz de Alarcón 13, 28014 Madrid, Spain
Dock House, The Waterfront, Cape Town 8001, South Africa

http://www.cambridge.org

First published 1966
First paperback edition 2002

Library of Congress Catalogue Card Number: 66-13387

ISBN 0 521 04869 9 hardback
ISBN 0 521 52336 2 paperback

CONTENTS

Chapter 4. *S*-matrix theory

PREFACE

One of the most remarkable discoveries in elementary particle physics has been that of the existence of the complex plane. From the early days of dispersion relations onwards the analytic approach to the subject has proved a most useful tool. In this monograph we give an account of two aspects of this approach. The first topic, dealt with in chapters 2 and 3, is the study of perturbation theory as a model with heuristic value as a guide to what we may expect to be true in the actual physical theory. The second topic, dealt with in chapter 4, is the exploration of what we may hope will prove to be the actual physical theory, that is to say the investigation of the properties of the analytic S-matrix. Our concern is to explain the principles as far as they are currently understood. For that reason we give no detailed account of the comparison, where it is possible, of prediction with experiment, nor of the various bootstrap models which have been constructed and which seem to offer the exciting possibility of a synthesis between the analytic approach and the other great weapon in the particle physicist's armoury, the group-theoretic exploitation of symmetries. Accounts of these vital topics can be found elsewhere.

Chapter 1 is intended to set the scene and to introduce in an elementary way various notions which constitute a background to what follows. Chapter 2 develops the analytic properties of Feynman integrals. The Landau equations and Cutkosky discontinuity formula are ideas encountered here which prove more general than their perturbation theory origin. Chapter 3 discusses asymptotic behaviour. This chapter is in principle separate from the material of chapters 2 and 4, though in practice a knowledge of high-energy behaviour is necessary in any computational scheme that uses dispersion relations. Chapter 4 is concerned with S-matrix theory. The earlier sections discuss various basic properties of the S-matrix and should be intelligible to a student who has read only chapter 1. From §4.10 onwards, however, a knowledge of at least the material of §§2.1–2.3, 2.9, 2.10 is desirable.

We are grateful to M. J. W. Bloxham, P. R. Graves-Morris, P. Osborne, T. W. Rogers, A. R. Swift and M. O. Taha for reading various parts of the manuscript and to Mrs Pamela Landshoff for typing most of it. We would like also to thank the staff of the

Cambridge University Press for their help and care in the preparation and printing of this book. We apologise to those authors who fail to find their names in the list of references. We have tried to give due credit for the ideas we use. Perhaps it is a consolation to reflect that the psychologists tell us that unconscious influences are the most powerful.

R. J. E.
P. V. L.
D. I. O.
J. C. P.

Cambridge
July 1965

CHAPTER 1

INTRODUCTION

1.1 Survey of objectives

We begin by considering the motivation for developing an S-matrix theory of particle interactions. The purpose of the book is to indicate how such a theory may be developed from physical principles and to discuss some of the properties of the S-matrix. Particular attention will be given to its analyticity properties and our study of these will rest very largely, though not entirely, on an analysis of the corresponding properties of Feynman integrals.

At present it is believed that the forces between particles fall into four categories, depending on their strength. The most familiar of these is the electromagnetic force, which has been expressed in terms of a field since the work of Maxwell. The quantisation of the electromagnetic field finally resolved the old paradox of the wave and particle nature of light.

The quantisation procedure uses either a Lagrangian or a Hamiltonian whose form is taken from classical physics. A solution of the resulting equations can be achieved in the form of a perturbation series expansion in powers of the square of the electric charge which, in rationalised units, is

$$e^2 = \tfrac{1}{137}.$$

The two most important difficulties encountered in this perturbation solution arise from two types of divergence. One of these, the infra-red divergence, can be eliminated in principle by taking account of the fact that the zero mass of the photon makes it impossible for the number of zero-energy photons to be measured. The other, the ultra-violet divergence, is eliminated by renormalisation, although it may be thought that the manipulation of infinite constants is still an unsatisfactory feature of the theory. For an account of these methods the reader is referred to books on quantum field theory, for example Schweber (1961); here we remark only that the results are in very good agreement with experiment.

The type of interaction with which S-matrix theory is mainly concerned comprises all strong interactions (we do not distinguish between these and the possible 'super-strong' interactions). These are responsible for nuclear forces and for the production of strange particles. Earlier formulations of a theory of strong interactions have proceeded analogously to electromagnetic theory. In the simplest form the strong interactions correspond to a field that is carried by the π-meson just as the electromagnetic field is carried by the photon. More generally a formal theory can be set up that involves the fields of all strongly interacting particles. However, there is a serious obstacle to the solution of the resulting equations since the only known methods of solution are based on a perturbation series in powers of the coupling constant and, in dimensionless units, so as to compare it with e^2, the square of this constant has a value about 15. Thus the perturbation series does not even begin to converge and a solution based on the first few terms is very unlikely to be useful.

In the last ten years a new approach to strong interactions has been developed which avoids the obvious defect of an expansion in the coupling constant, based on field theory. It is recognised that the fields themselves are of little interest, but that they are merely used to calculate transition amplitudes for interactions. These amplitudes are the elements of the S-matrix. The new approach is concerned with a direct study of the S-matrix, without the introduction of fields. It was first suggested by Heisenberg much earlier (Heisenberg, 1943; see also Møller, 1945, 1946) that the S-matrix might provide a means of avoiding the divergence difficulties of field theory, which at that time had not been solved by renormalisation. Heisenberg's formulation of S-matrix theory is in spirit very close to the formulation of a deductive S-matrix theory which will be described in chapter 4 of this book. However at that time (1943–52) progress was much more difficult, because a knowledge of the analytic properties of perturbation theory was not available to provide the guide-lines for applications of the S-matrix theory and for the formation of a deductive theory. The main parts of chapters 2 and 3 of this book will be concerned with the analytic properties of perturbation theory.

The deductive approach to S-matrix theory is based on the idea that one should try to calculate S-matrix elements directly, without the use of field quantities, by requiring them to have some general properties that ought to be valid, whether or not some underlying Lagrangian theory exists. (There is a tendency in the literature to call these properties 'axioms', but we do not use this term since it would suggest a degree of mathematical rigour that is lacking in the present state of the subject.) A list of the important properties to be satisfied by the S-matrix would include:

    (a) the superposition principle of quantum mechanics;

    (b) the requirements of special relativity;

    (c) the conservation of probability;

    (d) the short-range character of the forces;

    (e) causality and the existence of macroscopic time.

Notice that (d) actually excludes the electromagnetic interaction and there is at present no S-matrix theory which properly includes the presence of photons. The essential difficulty is the same one as leads to the infra-red divergence of perturbation theory, that the number of massless particles is not measurable. In practice one uses a combination of a perturbation series for the electromagnetic interaction and S-matrix theory for the strong interactions, but this procedure does not overcome the difficulty of principle where photons are involved.

The property (e) is the one whose consequences are most difficult to derive rigorously, and at the same time it is one of the most important. It is generally believed that the causality property requires the transition amplitudes to be the real-boundary values of analytic functions of complex variables. In view of the difficulties in deriving this result rigorously it is common to replace the property (e) by the assumption

    (e') transition amplitudes are the real-boundary values of
          analytic functions.

This assumption is much more precise mathematically than the property of causality but its physical meaning is more obscure. We will illustrate the connections between the two, and indicate the nature of the difficulty of making it rigorous, by considering a simple example.

Let $A(z,t)$ be a wave packet travelling along the z-direction with velocity v:

$$A(z,t) = \frac{1}{(2\pi)^{\frac{1}{2}}} \int_{-\infty}^{\infty} d\omega\, a(\omega) \exp\left\{i\omega\left(\frac{z}{v}-t\right)\right\}. \qquad (1.1.1)$$

Suppose this wave packet is scattered by a particle fixed at the origin $z = 0$. The scattered wave, in the forward direction, may be written as

$$G(r,t) = \frac{1}{r(2\pi)^{\frac{1}{2}}} \int_{-\infty}^{\infty} d\omega\, f(\omega)\, a(\omega) \exp\left\{i\omega\left(\frac{z}{v}-t\right)\right\}. \qquad (1.1.2)$$

The inverse of equation (1.1.1) is

$$a(\omega) = \frac{1}{(2\pi)^{\frac{1}{2}}} \int_{-\infty}^{\infty} dt\, A(0,t) \exp(i\omega t), \qquad (1.1.3)$$

and this tells us that if the incident wave does not reach the scatterer before time $t = 0$, that is if

$$A(0,t) = 0 \quad \text{for} \quad t < 0, \qquad (1.1.4)$$

then $a(\omega)$ is regular in the upper half of the plane of the variable ω now regarded as complex. For if, as we assume, the integral (1.1.3) converges for real ω, it will, by virtue of (1.1.4), converge even better for $\text{Im}(\omega) > 0$.

We now impose a causality condition,

$$G(r,t) = 0 \quad \text{for} \quad vt - r < 0. \qquad (1.1.5)$$

This expresses the requirement that no scattered wave reaches a point at distance r before a time v/r after the incident wave first reaches the scatterer. Then from the inverse of equation (1.1.2), using (1.1.5) we find that the product $a(\omega)f(\omega)$ is analytic in $\text{Im}(\omega) > 0$. Hence the scattering amplitude $f(\omega)$ itself is analytic in $\text{Im}(\omega) > 0$, except possibly at zeros of $a(\omega)$.

Arguments such as this about causality and analyticity can be made in various branches of classical physics, particularly in the theory of dispersion in optics (for a review see Hamilton (1959)). Thus the approach to high-energy physics that we describe in this book is often known as the 'dispersion relation' approach.

The difficulty in making the above discussion rigorous arises from the condition (1.1.4), which cannot actually be realised. This condition would imply a precise localisation in time (microscopic time) for the incident wave packet. But in S-matrix theory the quantity that we wish to know precisely is the energy, since this is one of the essential variables on which transition amplitudes depend. If we compromise, in accordance with the uncertainty principle, and set up our theory with only a partial knowledge of time and a partial knowledge of energy, our conclusions about analyticity are less precise. These problems relating to the use of macroscopic time persist in a relativistic formulation. They will be considered in more detail in chapter 4.

The causality condition in quantum field theory is usually assumed to correspond to the commutativity of the field operators for space-like separation of their arguments

$$[\phi(x), \phi(x')] = 0 \quad \text{for} \quad (x - x')^2 < 0, \qquad (1.1.6)$$

where for four-vectors we write $x = (x_0, \mathbf{x})$ and use the metric

$$x^2 = x_0^2 - \mathbf{x}^2. \tag{1.1.7}$$

Only for the electromagnetic field is the field operator physically observable, so the condition (1.1.6) for any other field can only have an indirect relation to causality in physics. Even if this condition is accepted it is very difficult to make use of it to prove rigorously any analytic properties of transition amplitudes (see, for example Froissart (1964)), though an heuristic derivation can be given fairly simply (Gell-Mann, Goldberger & Thirring (1954)). Within the framework of quantum field theory, without using perturbation expansions, only very limited information about analyticity properties has been obtained.

If, however, the perturbation series for a transition amplitude is used as a means for obtaining analyticity properties, much more information becomes available. The procedure, which will be followed in chapter 2 of this book, is to examine the analytic properties of individual terms in the perturbation series. Although one does not believe the magnitude of the individual terms to be significant, it is hoped that their analytic properties will indicate the analytic properties of the transition amplitude itself, particularly when properties are derived that hold for every term in the series. In chapter 3 this method is extended to include some aspects (particularly asymptotic behaviour) of the analytic properties of partial infinite sums of series within the full perturbation series.

In this book our discussion of the analytic S-matrix is limited to strong interactions. One hopes that in time a method for dealing with massless particles can be found. In the meantime, apart from sum rules, which seem to have limited scope, it is necessary to incorporate electromagnetic effects by perturbation theory. A similar situation is met with the weak interactions, at least where neutrinos are involved, though there are difficulties of renormalisation. In most practical situations the weakness of the interactions (about 10^{-10} compared with electromagnetic $1/137$) permits the use of first-order perturbation theory. The fourth category of force, the gravitational force, has a strength of order 10^{-40}. It again is believed to be transferred by a massless particle, the graviton, so it cannot at present be incorporated into a dispersion approach.

1.2   The S-matrix and its unitary and kinematic properties

In using the S-matrix to describe a scattering experiment we will assume that the forces are of sufficiently short range that the initial and final states consist effectively of free particles. These states can then be specified by the momentum of each particle together with certain discrete quantum numbers such as the spin and isospin. Due to the finite size of any experiment there is some residual uncertainty in the momentum but we assume that this is unimportant in practice. The momentum eigenvalues form a continuous spectrum but for clarity of notation in this section we will begin by using a discrete symbol m, or n, to label the states.

Let $|n\rangle$ denote the initial state of two particles that subsequently come together, interact, and separate. The superposition principle in quantum mechanics tells us that the final state can be written $S|n\rangle$, where S is a linear operator. The probability that a measurement on the final state gives a result corresponding to the state $|m\rangle$ is obtained from the square of the modulus of the matrix element

$$\langle m|S|n\rangle. \tag{1.2.1}$$

The set of states $|n\rangle$ is assumed to be orthonormal and complete,

$$\langle m|n\rangle = \delta_{mn}, \quad \sum_m |m\rangle\langle m| = 1. \tag{1.2.2}$$

Thus any state can be expressed by a superposition of the states $|n\rangle$, and the quantum numbers denoted by n uniquely specify a state.

If the initial state in a scattering experiment is the normalised state $|\rangle$, the total probability of the system ending up in some other state must be unity. Hence, writing

$$|\rangle = \sum_n a_n|n\rangle, \tag{1.2.3}$$

where $\sum_n |a_n|^2 = 1$, we obtain

$$1 = \sum_m |\langle m|S|\rangle|^2 = \sum_m \langle |S^\dagger|m\rangle\langle m|S|\rangle$$

$$= \langle |S^\dagger S|\rangle = \sum_{n,n'} a_{n'}^* a_n \langle n'|S^\dagger S|n\rangle. \tag{1.2.4}$$

In order for this to hold for all choices of the a_n, it is necessary that

$$\langle n'|S^\dagger S|n\rangle = \delta_{n'n},$$

or
$$S^\dagger S = 1. \tag{1.2.5a}$$

In the same way, the condition that the total probability be unity for an arbitrary final state to arise from some initial state gives

$$SS^\dagger = 1. \tag{1.2.5b}$$

Thus the operator S is unitary.

We consider next the consequences of relativistic invariance. If L is any proper Lorentz transformation, and if

$$L|m\rangle = |m'\rangle, \tag{1.2.6}$$

we require that $\qquad |\langle m'|S|n'\rangle|^2 = |\langle m|S|n\rangle|^2. \tag{1.2.7}$

in order that observable quantities be independent of the Lorentz frame. The definition of the S-matrix elements given above does not specify the phase uniquely. This permits us to replace (1.2.7) by the stronger condition

$$\langle m'|S|n'\rangle = \langle m|S|n\rangle. \tag{1.2.8}$$

For spinless particles this has the consequence that the matrix elements depend on the four-momenta only through their invariant scalar products. For example, the two-particle $\to$ two-particle matrix element

$$\langle p_3, p_4|S|p_1, p_2\rangle, \tag{1.2.9}$$

after removal of δ-functions specifying total energy-momentum conservation, can for the case of spinless particles be written as a function of the variables s, t, u, where

$$s = (p_1 + p_2)^2, \quad t = (p_1 - p_4)^2, \quad u = (p_1 - p_3)^2. \tag{1.2.10}$$

Notice that as a consequence of total energy-momentum conservation and the mass shell condition for each particle, these variables are not independent. From

$$p_1 + p_2 = p_3 + p_4, \quad p_i^2 = m_i^2 \quad (i = 1, 2, 3, 4), \tag{1.2.11}$$

it follows that $\qquad s + t + u = \sum_{i=1}^{4} m_i^2. \tag{1.2.12}$

The above form for the matrix elements applies only to spinless particles. However, even for these the elements of the S-matrix itself cannot be analytic, due to the occurrence of Dirac δ-functions. These occur in two ways. First, due to overall energy-momentum conservation, the matrix element (1.2.9) will contain a factor

$$\delta^{(4)}(p_1 + p_2 - p_3 - p_4).$$

Secondly, since the state-vectors occurring in (1.2.9) are momentum eigenstates they can contain no information about the positions of the particles in space. Hence they are overwhelmingly likely to be widely separated in space and not interact at all. When this happens, the four-momentum for each particle remains unchanged, and the S-matrix can therefore be separated usefully into two parts by writing (see Møller, 1945, 1946),

$$S = 1 + iR. \tag{1.2.13}$$

The relation between matrix elements of R and experimental cross-sections depends on the choice of normalisation for the free-particle states. These free-particle states are fully specified for spinless particles when the three-momentum of each particle is given, since the fourth component of p satisfies

$$p_0^2 = m^2 + \mathbf{p}^2. \tag{1.2.14}$$

We choose a covariant normalisation for the free particle states, so that the orthogonality and completeness relations, written symbolically in (1.2.2), become

$$\langle p'|p \rangle = (2\pi)^3 . 2p_0 \delta^{(3)}(\mathbf{p}' - \mathbf{p}), \tag{1.2.15}$$

$$\int |p'\rangle \frac{d^3\mathbf{p}'}{2p_0'(2\pi)^3} \langle p'|p \rangle = |p\rangle, \tag{1.2.16}$$

or equivalently

$$\int |p'\rangle \frac{d^4p' \delta^{(+)}(p'^2 - m^2)}{(2\pi)^3} \langle p'|p \rangle = |p\rangle. \tag{1.2.17}$$

Then the R-matrix element for two-particle scattering is related to a transition amplitude F by

$$\langle p_3, p_4|R|p_1, p_2 \rangle = (2\pi)^4 \delta^{(4)}(p_1 + p_2 - p_3 - p_4) F. \tag{1.2.18}$$

The cross-section is obtained from $|R|^2$ with integration over all possible final states when the incident flux is normalised to unity. For two-particle scattering this gives for the cross-section σ,

$$\sigma = \frac{1}{(8\pi)^2 qW} \int d\Omega |F|^2 \frac{p}{W}, \tag{1.2.19}$$

where $\mathbf{q}$ is the centre of mass momentum for a particle in the initial state, $\mathbf{p}$ for the final state, W is the centre of mass energy, and Ω is the solid angle in the final state. The differential cross-section for scattering to an angle (θ, ϕ) in the centre-of-mass system is

$$\frac{d\sigma}{d\Omega} = \frac{p}{(8\pi)^2 qW^2} |F|^2, \tag{1.2.20}$$

with $d\Omega = \sin\theta \, d\theta \, d\phi$.

The transition amplitude F is the amplitude that is given by a series of Feynman integrals when the particles have spin zero. It is these Feynman amplitudes that form the main part of our discussion of analyticity in the remaining sections of this chapter and the whole of chapter 2.

If the scattered particles have spin the discussion of analytic properties does not apply directly to the scattering amplitudes, and in addition our remarks about Lorentz invariance made earlier in this section for spinless particles must be modified. We will do no more than illustrate the differences that arise, and for a more complete discussion of the scattering of particles with spin and charge, the reader is referred to the account by Jacob (Chew & Jacob, 1964).

For our example we consider pion-nucleon scattering (Chew, Goldberger, Low & Nambu, 1957). There the amplitude F that occurs in (1.2.18) can be expressed in the form

$$F = 2m\bar{u}(p_3)\{A - \tfrac{1}{2}iB\gamma_\mu(p_2^\mu + p_4^\mu)\}u(p_1), \qquad (1.2.21)$$

where u denotes the Dirac spinor for the nucleon lines, for which p_3 and p_1 are the four-momenta. The quantities A and B are functions of two independent invariants chosen from s, t, u given by equations (1.2.10). The analytic properties of A and B are essentially the same as those of the amplitude F for scalar particles, to each order in perturbation theory. The other factors in equation (1.2.21) are often referred to as 'inessential complications'. Needless to say they are crucial in establishing relations between analyticity and experimental results.

It is frequently convenient in pion-nucleon scattering to make also a separation of the transition amplitude F into isospin amplitudes $F(\tfrac{1}{2})$ and $F(\tfrac{3}{2})$,

$$\left.\begin{array}{l} F(\pi^+ p \to \pi^+ p) = F(\tfrac{3}{2}), \\[4pt] F(\pi^- p \to \pi^- p) = \tfrac{1}{3}F(\tfrac{3}{2}) + \tfrac{2}{3}F(\tfrac{1}{2}), \\[4pt] F(\pi^- p \to \pi^0 n) = \tfrac{1}{3}\sqrt{2}\{F(\tfrac{3}{2}) - F(\tfrac{1}{2})\}. \end{array}\right\} \qquad (1.2.22)$$

In this book we are concerned with analytic properties of the S-matrix and not with the important considerations about spin and isospin which have been fully described elsewhere (Chew & Jacob, 1964). For most of our discussion we will therefore consider only the interactions of particles that have zero spin and isospin. Then the amplitude F is given by the series of Feynman integrals whose analyticity we will describe in chapters 2 and 3.

For the case of elastic scattering of two spinless particles a consequence of Lorentz invariance is the symmetry of the matrix element:

$$\langle m|S|n\rangle = \langle n|S|m\rangle, \qquad (1.2.23)$$

that is $\langle p_3, p_4|S|p_1, p_2\rangle = \langle p_1, p_2|S|p_3, p_4\rangle.$

This is because in the centre-of-mass system (the Lorentz frame in
which $\mathbf{p}_1 + \mathbf{p}_2 = 0 = \mathbf{p}_3 + \mathbf{p}_4$) a rotation of π about the bisector of the
angle between $\mathbf{p}_1$ and $\mathbf{p}_3$ interchanges these momenta, and it also
interchanges $\mathbf{p}_2$ and $\mathbf{p}_4$ (see Fig. 1.2.1). So (1.2.23) follows as a result

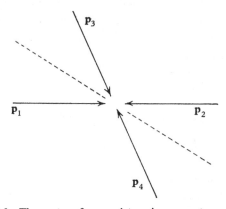

Fig. 1.2.1. The centre-of-mass picture in momentum space for
two-particle scattering $1 + 2 \rightarrow 3 + 4$.

of (1.2.8). It need not, however, be true for other amplitudes, though
it sometimes can be deduced from the invariance of strong interactions
under the operation PT. This is true for two-particle $\rightarrow$ two-particle
amplitudes.

1.3 Analyticity, crossing and dispersion relations

We now discuss in more detail some of the analyticity properties of
the scattering amplitudes that will be derived in chapter 2. First we
consider some consequences of unitarity.

Substitute from (1.2.13) into the unitarity relation (1.2.5), giving

$$R - R^\dagger = iR^\dagger R = iRR^\dagger, \qquad (1.3.1)$$

or, in the notation of section (1.2), for two-particle scattering

$$\langle p_3, p_4|R|p_1, p_2\rangle - \langle p_1, p_2|R|p_3, p_4\rangle^* = i\langle p_3, p_4|R^\dagger R|p_1, p_2\rangle \qquad (1.3.2a)$$
$$= i\langle p_3, p_4|RR^\dagger|p_1, p_2\rangle, \qquad (1.3.2b)$$

where the star denotes complex conjugation, and the dagger hermitian conjugation. If the symmetry condition (1.2.23) is valid the left-hand side of (1.3.2) is just twice the imaginary part of the matrix element:

$$2i \, \mathrm{Im} \langle p_3, p_4 | \, R \, | p_1, p_2 \rangle. \tag{1.3.3}$$

Then the unitary relation (1.3.2) becomes

$$2 \, \mathrm{Im} \langle p_3, p_4 | \, R \, | p_1, p_2 \rangle = \sum_n \langle n | \, R \, | p_3, p_4 \rangle^* \langle n | \, R \, | p_1, p_2 \rangle \tag{1.3.4a}$$

$$= \sum_n \langle p_3, p_4 | \, R \, | n \rangle \langle p_1, p_2 | \, R \, | n \rangle^*, \tag{1.3.4b}$$

where the Σ denotes a sum and an integral over all intermediate states that are allowed by conservation of the total energy and momentum. Thus for total energies below the inelastic threshold the unitarity condition is, in terms of the amplitude F of (1.2.18)

$$2 \, \mathrm{Im} \langle p_3 p_4 | \, F \, | p_1 p_2 \rangle = (2\pi)^{-2} \int \frac{d^3 \mathbf{k}_1 d^3 \mathbf{k}_2}{W^2} \delta^{(4)}(p_1 + p_2 - k_1 - k_2)$$

$$\times \langle p_3 p_4 | \, F \, | k_1 k_2 \rangle \langle p_1 p_2 | \, F \, | k_1 k_2 \rangle^*$$

$$= (2\pi)^{-2} \int d^4 k_1 \, d^4 k_2 \delta^{(+)}(k_1^2 - m^2) \, \delta^{(+)}(k_2^2 - m^2)$$

$$\times \delta^{(4)}(p_1 + p_2 - k_1 - k_2) \langle p_3 p_4 | \, F \, | k_1 k_2 \rangle$$

$$\times \langle p_1 p_2 | \, F \, | k_1 k_2 \rangle^*, \tag{1.3.5}$$

where W is the centre of mass energy.

Above the energy-threshold for inelastic scattering a new term must be added to the right-hand side of the unitarity relation (1.3.5) so as to include the extra intermediate states that are allowed by energy-conservation. This implies a change in the left-hand side, and suggests that the elastic scattering matrix-element has a singularity at each energy corresponding to a threshold for a new allowed physical process. This is our first encounter with an effect of unitarity on analyticity of the S-matrix; later, in chapter 4, we will consider these effects in more generality.

The thresholds are branch-points of the amplitude F (Eden, 1952), as we shall see in chapter 2, so we draw cuts in the complex energy-squared plane ($s = W^2$), attached to the branch-points and by convention running along the real axis. The purpose of the cuts is to make the amplitude single valued on a Riemann surface. If we do not cross the cuts in Fig. 1.3.1, we have a single sheet of this Riemann surface. This is called the 'physical sheet' if the physical scattering amplitude is a boundary value on the real cut of the amplitude on this sheet.

Other sheets of the Riemann surface, associated with the amplitude F as a function of s, are reached by burrowing through a branch cut or through several branch cuts, to reach another layer of this multi-layer surface. These other sheets are called unphysical sheets and they are to be distinguished from each other by the manner in which they are connected to the physical sheet, for example by specifying which branch cuts must be crossed to reach the physical sheet.

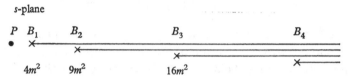

Fig. 1.3.1. Branch cuts for the scattering amplitude F in the complex s-plane arising from normal thresholds B_1, B_2, ... The point P denotes a pole in F.

For two-particle scattering, the elastic amplitude F is a function of two variables which, to be definite, we choose as s, t given by (1.2.10),

$$\langle p_3 p_4 | F | p_1 p_2 \rangle = F(s,t). \qquad (1.3.6)$$

We have so far been considering F as a function of the invariant energy-squared s, keeping the momentum-transfer-squared t, fixed. The branch-points shown in Fig. 1.3.1 at $s = 4m^2$, $9m^2$, $16m^2$ are called 'normal thresholds' and correspond to the energies at which production of extra particles is possible. The leading normal threshold, $s = 4m^2$, is the least energy-squared at which a two-particle state can exist. We assume not only equal-mass particles in this example, but also that no conservation law except energy precludes the creation of any particular number of particles from a two-particle state. With this assumption, conservation laws (excluding energy) do not forbid going from a two-particle state to a one-particle state. It is assumed that such a state corresponds to a singularity of the amplitude $F(s,t)$, reached at an unphysical value of the variable s below the leading normal threshold, at

$$s = m^2. \qquad (1.3.7)$$

This singularity is denoted P in Fig. 1.3.1, and in perturbation theory it is a pole, not a branch-point. Using unitarity and causality we will show more generally in § 4.5 that the singularity must be a pole.

The region in which $F(s,t)$ is the amplitude for the physical scattering process

$$A_1 + A_2 \to A_3 + A_4 \qquad (1.3.8)$$

must have real positive energy $p_i^{(0)}$ for each particle, and real three-momentum $\mathbf{p}_i$. In the equal-mass case this gives

$$s \geqslant 4m^2, \quad t \leqslant 0, \quad u \leqslant 0. \tag{1.3.9}$$

This result can be obtained by expressing s, t, u in terms of the momentum $\mathbf{q}$ and the scattering angle θ in the centre-of-mass system, which gives

$$\left.\begin{aligned}
s &= 4(m^2 + q^2), \\
t &= -2q^2(1 - \cos\theta), \\
u &= -2q^2(1 + \cos\theta).
\end{aligned}\right\} \tag{1.3.10}$$

When the masses are not equal the conditions are not quite so simple; they are derived by Kibble (1960) (see also §4.3).

So far we have varied only s in discussing analytic continuation, but in general both s and t can be regarded as complex variables in the amplitude $F(s, t)$. Then we can consider analytic continuation from the physical region (1.3.9) to the region

$$u \geqslant 4m^2, \quad s \leqslant 0, \quad t \leqslant 0. \tag{1.3.11}$$

It is assumed that the resultant function F, evaluated in a suitable limit on to the region (1.3.11), is the physical scattering amplitude for the process

$$A_1 + \bar{A}_3 \to \bar{A}_2 + A_4, \tag{1.3.12}$$

where $\bar{A}_i$ denotes the anti-particle of A_i. For this process the energy in the centre-of-mass frame for the initial (or final) state is just $\sqrt{u}$. It is further assumed that by analytic continuation to the region

$$t \geqslant 4m^2, \quad u \leqslant 0, \quad s \leqslant 0, \tag{1.3.13}$$

the function F, evaluated in a suitable limit, gives the physical scattering amplitude for the process

$$A_1 + \bar{A}_4 \to \bar{A}_2 + A_3 \tag{1.3.14}$$

for which the energy in the centre-of-mass frame is $\sqrt{t}$.

These important properties are called the 'crossing' properties. They state that the same analytic function can be used to describe the three different physical processes (1.3.8), (1.3.12), (1.3.14) by making an appropriate choice of physical values for the variables s and t (or u). These physical processes are often called different 'channels', and one refers to them as the s-channel, the t-channel and the u-channel when

s, t and u, respectively, are the energy variables. Remembering the relation (1.2.12), which here becomes

$$s+t+u = 4m^2, \qquad (1.3.15)$$

we can draw the physical regions for the three channels using oblique axes as in Fig. 1.3.2 (Mandelstam, 1958).

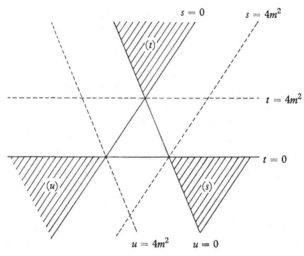

Fig. 1.3.2. The Mandelstam diagram using oblique axes showing the physical regions (shaded areas) in which s, or t, or u denotes the square of the centre-of-mass energy for equal-mass particles in collision.

Since we now have symmetry between the three variables s, t, u, it is convenient to change the signs of the four-momenta from those used in §1.2 (equation (1.2.10)) so as to give

$$\left.\begin{aligned} s &= (p_1+p_2)^2 = (p_3+p_4)^2, \\ t &= (p_1+p_4)^2 = (p_2+p_3)^2, \\ u &= (p_1+p_3)^2 = (p_2+p_4)^2. \end{aligned}\right\} \qquad (1.3.16)$$

This convention will often be used in the remainder of this book.

A further convention that is sometimes used in the literature is to write the amplitude $F(s,t,u)$ as a function of three variables, but with the constraint (1.3.15) relating s, t and u. In fact, F is defined only when (1.3.15) is satisfied so this formal achievement of symmetry is somewhat ambiguous, and in practice it is better to regard F as a function of two variables, $F(s,t)$ or $F(s,u)$, for example. Similarly, it is usually easier to work in the real s, t-plane with orthogonal axes, rather than with the oblique axes shown in Fig. 1.3.2. We then obtain Fig. 1.3.3 which illustrates the physical regions for the three processes (1.3.8), (1.3.12) and (1.3.14).

If we commence from the u-channel, with process (1.3.12), and cross the particles A_1 and A_4 we obtain the process

$$\bar{A}_4 + \bar{A}_3 \to \bar{A}_1 + \bar{A}_2. \tag{1.3.17}$$

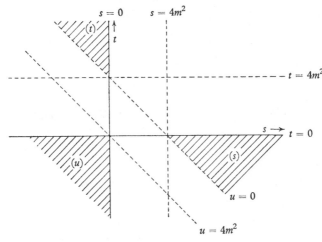

Fig. 1.3.3. Physical regions (shaded areas) for equal-mass scattering shown in the real (s, t)-plane.

The physical region for this process involving anti-particles is the same as that for the process (1.3.8) involving particles. The TCP theorem asserts that the amplitudes for these two processes are the same. Its proof is discussed in § 4.8.

The reactions (1.3.12) and (1.3.14) also have TCP-inverses,

$$\left. \begin{array}{l} A_2 + \bar{A}_4 \to \bar{A}_1 + A_3, \\ A_2 + \bar{A}_3 \to \bar{A}_1 + A_4, \end{array} \right\} \tag{1.3.18}$$

so that altogether crossing and TCP relate the amplitudes for six physical processes to the same function $F(s, t)$. Combining this with the symmetry (1.2.23) [PT invariance] that is valid in strong interactions, we obtain a further six processes by reversing the direction of the above reactions.

Just as we were able to deduce the existence of singularities at the normal thresholds from unitarity, it is possible also to deduce the existence of further singularities from the assumption of crossing symmetry. Since $\sqrt{t}$ and $\sqrt{u}$ represent energies in the t and u channels, they will yield branch points exactly corresponding to those drawn in $B_1, B_2, B_3, \ldots$ for s in Fig. 1.3.1, that is at

$$t = 4m^2, \quad 9m^2, \quad 16m^2, \quad \ldots, \tag{1.3.19}$$

$$u = 4m^2, \quad 9m^2, \quad 16m^2, \quad \ldots. \tag{1.3.20}$$

If we fix, say, u at a real value u_0, then because of (1.3.15), the branch-points (1.3.19) will appear in the s-plane at

$$s = -u_0, \quad -u_0 - 5m^2, \quad -u_0 - 12m^2, \quad \ldots, \tag{1.3.21}$$

and the $t = m^2$ pole will appear at

$$s = -u_0 + 3m^2. \tag{1.3.22}$$

The resulting picture (for fixed $u = u_0$) in the complex s-plane is shown in Fig. 1.3.4. This figure depicts the physical sheet. As we remarked earlier this is called the physical sheet since the amplitude $F(s, u_0)$ becomes the physical amplitude for a suitably chosen value of s on this sheet. It is of course a matter of convention that the branch cuts are drawn along the real s-axis. Only their end-points are fixed and they can be distorted as desired without changing the value of the function F. The branch-points however are fixed and cannot be moved so long as the parameter u is kept fixed at u_0.

With real branch-cuts in the s-plane it is necessary to decide which limit on to the branch-cut gives the physical amplitude. For $u_0 < 0$, and with s real and $s > 4m^2 - u_0$, we have physical values of s, t, u that correspond to the s-channel shown in Fig. 1.3.2. We shall see in §2.3 that perturbation theory shows the physical amplitude to be given by the limit on to this right-hand cut from the upper-half s-plane,

$$F(\text{physical}) = \lim_{\epsilon \to 0+} F(s + i\epsilon, u_0). \tag{1.3.23}$$

This result is obtained by showing it to be equivalent to Feynman's prescription for obtaining physical amplitudes by giving a small negative imaginary part ($-i\epsilon$) to the mass of each particle in any internal line of a Feynman diagram. With this rule, each Feynman integral can be evaluated with real external four-momenta, that is real s.

This result (1.3.23) from perturbation theory is referred to as 'the $i\epsilon$-prescription'. Its derivation and significance outside the framework of perturbation theory is discussed in §4.4.

If, as is generally the case, the symmetry condition (1.2.23) is not valid, the expression appearing on the left-hand side of the unitarity condition is not the imaginary part of the amplitude. In this case, as is shown in §4.6, the amplitude that gives $\langle p_1 p_2 | R | p_3 p_4 \rangle^*$ is related to that which gives $\langle p_3 p_4 | R | p_1 p_2 \rangle$ only by analytic continuation. The latter amplitude, according to (1.3.23), is the limit on to the real axis of the complex s-plane from above and we show in §4.6 that the former is the limit, of the *same* analytic function, from below. This property is known as 'hermitian analyticity' (Olive, 1962). It has the consequence that the left-hand side of the unitarity relation (1.3.2) involves the *discontinuity* of the analytic function across the normal-threshold cuts.

When the symmetry relation is valid, the hermitian analyticity property results in the amplitude F being real on the part of the real axis between the branch cuts illustrated in Fig. 1.3.4. This is proved directly from the Schwartz reflection principle in the theory of functions of a complex variable.

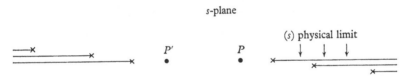

s-plane

Fig. 1.3.4. Branch cuts in the complex s-plane arising from thresholds in the s-channel (on the right), and in the t-channel (on the left) for a fixed value of u. Two poles are also shown.

We can use crossing symmetry to derive the physical amplitude for the t-channel from the function $F(s, u_0)$ whose singularities are shown in Fig. 1.3.4. With $u_0 < 0$, we need to go to a real value $t > 4m^2 - u_0$, and take the limit from the upper half t-plane, $t + i\epsilon \to t$. From the relation (1.3.15) this is equivalent to taking s from its physical value for $F(s, u_0)$ in the s-channel, through the upper half s-plane to the gap in the branch cuts along the real axis, across this gap into the lower half s-plane, and then taking the limit $F(s - i\epsilon, u_0)$ for $s < 0$. Thus two physical amplitudes can be obtained as boundary values of $F(s, u_0)$.

Dispersion relations

In the remainder of this section we will show how many of the simpler analyticity properties of a scattering amplitude can be summarised by 'dispersion relations'. This is the physicist's terminology for the Hilbert transforms that are well known to mathematicians.

Let us assume that the singularities shown in Fig. 1.3.4 represent all the singularities of $F(s, u_0)$ on the physical sheet. This assumption can be justified to every order in perturbation theory for a number of important physical situations, including the scattering of equal mass particles (for $-4m^2 < u_0 < 4m^2$), as we shall discuss in § 2.5.

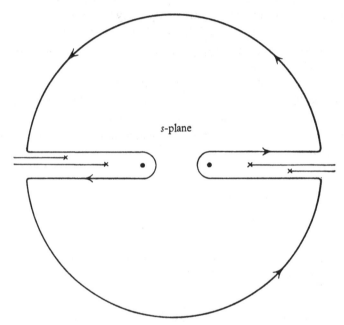

Fig. 1.3.5. The Cauchy contour for the amplitude F in the s-plane that is used for deriving a dispersion relation.

Let C be the contour drawn in Fig. 1.3.5, so that inside C the function $F(s, u_0)$ is regular. From Cauchy's theorem we obtain the result,

$$F(s, u_0) = \frac{1}{2\pi i} \int_C \frac{ds' F(s', u_0)}{s' - s}. \qquad (1.3.24)$$

Let us assume that $F(s', u_0) \to 0$ as $|s'| \to \infty$. Then the contribution to the integral (1.3.24) from the curved part of the contour C will tend to zero as we let its radius tend to infinity. This gives

$$F(s, u_0) = P + \frac{1}{2\pi i} \int_{-u_0}^{-\infty} \frac{ds' F_t(s', u_0)}{s' - s} + \frac{1}{2\pi i} \int_{4m^2}^{\infty} \frac{ds' F_s(s', u_0)}{s' - s}. \qquad (1.3.25)$$

Here F_s and F_t respectively represent the discontinuities of F across the right-hand and the left-hand cuts shown in Fig. 1.3.4, while P represents the contribution from the two poles in this figure,

$$P = \frac{g_s^2}{s - m^2} + \frac{g_t^2}{s + u_0 - 3m^2},\qquad (1.3.26)$$

where g_s and g_t are constant.

If we make use of (1.3.15) and define also

$$s' + t' + u = 4m^2, \qquad (1.3.27)$$

we can write (1.3.25) and (1.3.26) together in a more elegant form using the notation $F(s,t,u)$,

$$\begin{aligned}
F(s,t,u) = \frac{g_s^2}{s - m^2} + \frac{g_t^2}{t - m^2} &+ \frac{1}{2\pi i} \int_{4m^2}^{\infty} \frac{ds' F_s(s',t',u)}{s' - s} \\
&+ \frac{1}{2\pi i} \int_{4m^2}^{\infty} \frac{dt' F_t(s',t',u)}{t' - t}. \quad (1.3.28)
\end{aligned}$$

For historical reasons, as we noted earlier, this type of equation is called a 'dispersion relation'. The form (1.3.28) is a dispersion relation with u fixed (that is with u not integrated). Similar dispersion relations can be written down with either s or t fixed.

The dispersion relation (1.3.28), with u fixed, contains the discontinuities F_s and F_t respectively associated with the thresholds in the s- and t-channels. The other two dispersion relations involve also the discontinuity F_u in the u-channel. Because of the hermitian analyticity property these discontinuities are given by unitarity, and when the symmetry relation (1.2.23) holds they are also twice the imaginary part of F. In the literature the discontinuity of F and twice the imaginary part are often regarded as interchangeable concepts, but one should always confirm first that the symmetry relation (1.2.23) holds.

If this is the case, which we assume merely for simplicity, from

$$F_s = \operatorname{disc} F = 2i \operatorname{Im} F, \qquad (1.3.29)$$

we have an expression for F_s in terms of an integral over the product of F and F^*, given by the right-hand side of (1.3.5) for elastic scattering with

$$4m^2 \leqslant s < 9m^2, \quad t \leqslant 0, \quad u \leqslant 0. \qquad (1.3.30)$$

Explicitly, this gives

$$F_s(s,t) = \frac{i}{(2\pi)^2} \frac{(s - 4m^2)^{\frac{1}{2}}}{8s^{\frac{1}{2}}} \int d\Omega\, F(s,t')\, F^*(s,t''), \qquad (1.3.31)$$

when (1.3.30) holds, where t' relates to the square of the momentum transfer in the first factor of the integrand of (1.3.5) and t'' to the same for the second factor; thus both depend on the solid angle Ω.

Above the inelastic threshold, for example with

$$9m^2 \leqslant s < 16m^2, \quad t \leqslant 0, \quad u \leqslant 0, \qquad (1.3.32)$$

an extra integral must be added to the right-hand side of equation (1.3.31) so as to include the allowed three-particle intermediate states. Further terms must be added as s is increased past each threshold.

If t and u are not both negative in the dispersion relation (1.3.28), then F_s must be obtained by analytic continuation of the equation (1.3.31) from the physical region (or from the appropriate unitarity equation including inelastic terms). The other discontinuities F_t and F_u can be related to F and F^* in similar ways, using unitarity in the t and u channels respectively.

The combined set of three dispersion relations and three discontinuity relations (unitarity) is often called a dynamical set of equations. Since they are coupled equations for the amplitude it is clear that they impose severe restrictions on its form. Indeed, if (1.3.31) represented the full discontinuity relation, instead of just the part appropriate to elastic scattering, we would have a closed set of equations for $F(s, t)$. But, unfortunately, the discontinuity relations giving F_s in the inelastic regions, like (1.3.32), involve production amplitudes in addition to F itself. The system of dynamical equations cannot be closed without the addition of generalised dispersion relations for production amplitudes. The latter are very complicated, and in practice most 'dynamical' calculations have been based primarily on the elastic scattering approximation to these equations. Further details of this type of work in S-matrix theory are given by Chew (Chew & Jacob, 1964). It represents an important aspect of S-matrix theory for which it is essential to have as much information as possible about analyticity properties both of scattering amplitudes and of production amplitudes.

There is one particularly interesting extension of the dispersion relation (1.3.28), that provides a means for obtaining the analytic continuation of F_s which is required in the dynamical programme of S-matrix theory. This extension was proposed by Mandelstam (1958) and involves a double dispersion relation. It exploits the analyticity of $F(s, t)$ as a function of two complex variables, or $F(s, t, u)$ with the restriction (1.3.15).

The representation proposed by Mandelstam in a simple case has the form,

$$F(s,t,u) = P + \frac{1}{\pi^2}\iint \frac{\rho_{st}(s',t')}{(s'-s)(t'-t)}\,ds'\,dt'$$
$$+ \frac{1}{\pi^2}\iint \frac{\rho_{tu}(t',u')}{(t'-t)(u'-u)}\,dt'\,du' + \frac{1}{\pi^2}\iint \frac{\rho_{us}(s',u')}{(u'-u)(s'-s)}\,du'\,ds',$$

$$(1.3.33)$$

where P denotes the pole terms, and $s'+t'+u' = 4m^2$. This is believed to be valid for the scattering of spinless particles in an equal mass theory on the assumption that F tends to zero as the variables tend to infinity in any direction in the complex plane. It has however not been fully proved even within the framework of perturbation theory. Moreover, it seems that F does not have the required behaviour at infinity in the complex plane (this is discussed in § 3.8).

In the Mandelstam representation (1.3.33), the spectral function $\rho_{st}(s',t')$ represents a double discontinuity across cuts in the s-channel and the t-channel simultaneously; ρ_{tu} and ρ_{us} are defined similarly. The integrations are over certain real regions of the variables s', t', u'. The representation states that $F(s,t,u)$ is analytic in the topological product of the complex planes of s, t and u cut along their real axes and with the relation (1.3.15) between these variables.

In practice the greatest use of the Mandelstam representation has been in regions where one of the variables remains finite, and it gives simply an extension of a single variable dispersion relation by analytic continuation. Another valuable use has been in the derivation of partial-wave dispersion relations. Neither of these contain so much information about F as (1.3.33) and in some cases they can be proved for every term in perturbation theory.

The partial-wave dispersion relations are of considerable practical importance. We will consider their form for equal masses. Then if θ is the angle of scattering in the centre-of-mass frame in the s-channel

$$t = \tfrac{1}{2}(s-4m^2)(\cos\theta - 1). \qquad (1.3.34)$$

Substituting for t into F we obtain

$$\tilde{F}(s,\cos\theta) = F(s,t). \qquad (1.3.35)$$

The partial wave amplitude $a_l(s)$ is defined by

$$a_l(s) = \frac{1}{2}\int_{-1}^{1} d(\cos\theta)\,\tilde{F}(s,\cos\theta)\,P_l(\cos\theta). \qquad (1.3.36)$$

For physical values of s and $\cos \theta$ we obtain the usual series

$$\tilde{F}(s, \cos \theta) = \sum_{l=0}^{\infty} (2l+1)\, a_l(s)\, P_l(\cos \theta). \qquad (1.3.37)$$

A useful feature of this series is that it converges also when s is physical and $\cos \theta$ is complex but within a certain ellipse (Lehmann, 1958). This convergence follows from the analyticity of F within the same ellipse (Whittaker & Watson, 1940), which has been established in field theory by Lehmann, and can also be shown for each term in perturbation theory by the methods of chapter 2.

The value of the partial-wave series arises in practice from its use in approximation procedures. For scattering at low or medium energies a few partial waves may be adequate to approximate the full amplitude. A difficulty arises in analytic continuation, since the continuation of an approximate function may differ substantially from the continuation of the function itself. This obstacle has not been overcome in general.

Another useful aspect of partial wave amplitudes comes from the simple form of the unitarity condition, which becomes for elastic scattering,

$$\operatorname{Im} a_l(s) = \frac{1}{16\pi} \sqrt{\frac{s - 4m^2}{s}}\, |a_l(s)|^2. \qquad (1.3.38)$$

However, for inelastic scattering the difficulties from production processes mentioned earlier modify this relation in a way that has to be approximated in practical calculation. The price that is paid for obtaining such a simple form of unitarity is the loss of simplicity in the crossing relations. Since a partial wave refers to a definite channel in which s (or t or u) is the square of the energy, one obtains three types of partial amplitude with no very simple relation between them. A second penalty occurs even within a single channel. The integration in equation (1.3.36) that defines $a_l(s)$ leads to branch cuts in $a_l(s)$ that are not present in the full amplitudes. The simplest of these arises from the pole in one of the crossed channels and in the equal mass case leads only to an extension of the real left-hand branch-cut. In the unequal mass case the system of branch-cuts for partial waves may become quite complicated (MacDowell, 1959; Kennedy & Spearman, 1962).

1.4   High-energy behaviour and subtractions

The behaviour of a scattering amplitude at very high energies is important both for its effect on dispersion relations and also for its relation to resonances through the theory of Regge poles. The high-energy behaviour of amplitudes can be studied within the framework of perturbation theory by making partial sums of the perturbation series. These methods will be described in chapter 3. Here we will indicate briefly its relevance to dispersion relations and to Regge theory.

In order to derive the dispersion relation (1.3.25) we had to assume that

$$F(s, u_0) \to 0 \qquad (1.4.1)$$

as $|s| \to \infty$ with u_0 fixed. Suppose, however, that this condition does not hold, but that there exists an integer N such that

$$|F(s, u_0)| \sim \text{constant} \times |s|^{N-1} \qquad (1.4.2)$$

as $|s| \to \infty$. Then we can still obtain a dispersion relation when F is analytic in the cut s-plane.

Let $s_1, s_2, \ldots, s_N$ be constant. Then the function

$$G(s, u_0) = \frac{F}{(s-s_1)(s-s_2)\ldots(s-s_N)} \qquad (1.4.3)$$

will have the behaviour (1.4.1). We can therefore apply Cauchy's theorem to G, as we did in deriving the dispersion relation for F, and we can neglect the contribution from the circular part of the contour as its radius tends to infinity (see Fig. 1.3.5). However, now the function G has poles at $s_1, \ldots, s_N$ and we will therefore collect the residues at these poles. This leads to a dispersion relation for F in the form

$$F(s, t, u_0) = \phi^{(N)} + \frac{g_s^2}{s-m^2} + \frac{g_t^2}{t-m^2} + \frac{(s-s_1)(s-s_2)\ldots(s-s_N)}{2\pi i}$$

$$\times \left\{ \int_{4m^2}^{\infty} \frac{ds'\, F_s(s', t', u_0)}{(s'-s)(s'-s_1)\ldots(s'-s_N)} \right.$$

$$\left. + \int_{4m^2}^{\infty} \frac{dt'\, F_t(s', t', u_0)}{(t'-t)(t'-t_1)\ldots(t'-t_N)} \right\}, \qquad (1.4.4)$$

where the N quantities $t_1, \ldots, t_N$, are defined by

$$s_i + t_i + u_0 = 4m^2, \qquad (1.4.5)$$

and also

$$s' + t' + u_0 = 4m^2, \qquad (1.4.6)$$

so that as $|t'| \to \infty$, so also does $|s'|$, giving the behaviour (1.4.2).

The function $\phi^{(N)}$ in (1.4.4) is a polynomial of degree $(N-1)$ in s. Its N coefficients could only be found from a knowledge of the values of F at the N points s_i. These unknown coefficients are called 'sub-traction constants' and (1.4.4) is called a dispersion relation 'with N subtractions'.

It is therefore an important question whether such an integer N exists and, if it does, what its value is. The situation that is believed to hold is that N is a function of u_0. Thus for different fixed values of u we will require different numbers of subtractions. This will be discussed further in § 3.2.

A more general result is required in order to satisfy the double dispersion relation (1.3.33) suggested by Mandelstam. For the form given in (1.3.33) it was necessary to assume that

$$F(s, t, u) \to 0$$

as the variables s, t (with $u = 4m^2 - s - t$) tend to infinity simulta-neously (or separately) in any direction in their respective complex planes. If this is not the case, then it may be that F is bounded by a polynomial in the two variables s and t. In that case a dispersion relation like (1.3.33) could be written down together with subtrac-tions. For the validity of the Mandelstam representation with sub-tractions it is necessary that the number of subtractions $N(u_0)$ in the single variable dispersion relation (1.4.4) shall remain finite as $u_0 \to \infty$. However, a discussion of this question by Mandelstam (1963) suggests that in fact $N \to \infty$ as $u_0 \to \infty$. Then the double dispersion relation, even with subtractions, would not hold. The methods used in studying this problem are closely linked with Regge-theory using complex angular momenta.

Regge developed the use of complex angular momentum in non-relativistic potential scattering. For this it can be shown that an analytic function of two complex variables, l and s,

$$a(l, s) \tag{1.4.7}$$

can be defined so that it is equal to the partial wave amplitude $a_l(s)$ when l takes integer values (see (1.3.36)). It is then possible to replace the usual partial wave-series expansion (1.3.37) by an integral, the Watson–Sommerfeld transform, namely

$$\tilde{F}(s, \cos\theta) = \tfrac{1}{2}i \int_C \frac{dl(2l+1)\, a(l, s)\, P_l(-\cos\theta)}{\sin l\pi}, \tag{1.4.8}$$

where C is a contour in the complex l-plane that lies close to and surrounds the positive real axis.

If $a(l, s)$ can be defined so that it tends to zero suitably at infinity in the l-plane, which is the case for potential scattering, the contour C

can be opened out so that it lies parallel to the imaginary l-axis. In doing this one will collect residues at the poles l_n of $a(l, s)$ that lie in the right half l-plane. This gives

$$\tilde{F}(s, \cos \theta) = \sum_n (2l_n + 1) b_n \operatorname{cosec} (l_n \pi) P_{l_n}(-\cos \theta)$$

$$+ \tfrac{1}{2} i \int_{L-i\infty}^{L+i\infty} \frac{dl(2l+1)\, a(l, s)\, P_l(-\cos \theta)}{\sin l\pi}, \quad (1.4.9)$$

where b_n denotes the residue of a at the pole $l = l_n$.

This formula suggests that because

$$P_l(-\cos \theta) \sim (-\cos \theta)^l, \qquad (1.4.10)$$

the behaviour of the amplitude $\tilde{F}$, as $\cos \theta \to \infty$, will be dominated by the term in (1.4.9) that has the largest value of real l_n (note the integral lies parallel to $\operatorname{Re}(l) = 0$). Since, from (1.3.34), t is linearly proportional to $\cos \theta$, this conclusion can be written

$$F(s, t) \sim \text{constant} \times t^{\alpha'}, \qquad (1.4.11)$$

where α' denotes the real part of the pole l_n of $a(l, s)$ that lies farthest to the right in the complex l-plane. Clearly α' is a function of s in general,

$$\alpha' = \alpha'(s). \qquad (1.4.12)$$

In the t-channel, the argument we have outlined indicates that the asymptotic behaviour for large t depends on s as we stated above.

The poles of the partial amplitude $a(l, s)$ in the complex l-plane are called 'Regge' poles. We have assumed in the above discussion that $a(l, s)$ is meromorphic in the right-half l-plane. In fact, the work of Mandelstam (1963) and Polkinghorne (1963 d) indicates the existence of cuts in a crossing-symmetric relativistic theory, although meromorphy holds in potential theory with Yukawa potentials. These difficulties and a fuller discussion of Regge theory will be given in chapter 3 (see also Squires, 1963; Frautschi, 1963).

For real physical values of s and $\cos \theta$, Froissart (1961 b) and Martin (1964) have established bounds on the scattering amplitude. Their assumptions† include unitarity, and analyticity of the amplitude in the product of the cut physical sheets in the variables s, t, u. The results are, for real $s \to \infty$,

$$|\tilde{F}(s, \cos \theta = 1)| < \text{constant}\, s\, (\log s)^2, \qquad (1.4.13)$$

$$|\tilde{F}(s, |\cos \theta| < 1)| < \text{constant}\, \frac{s^{\frac{3}{4}} (\log s)^{\frac{3}{2}}}{(\sin \theta)^{\frac{1}{2}}}. \qquad (1.4.14)$$

---

† *Note added in proof.* Martin (1965) has recently derived these bounds using only analytic properties that have been established from axiomatic quantum field theory.

Although these results are not nearly enough to decide the question of subtractions, they do provide an important test for any assumptions that are made about Regge poles in a relativistic theory. In principle they can also be tested experimentally.

By means of a theorem known as the 'optical theorem' (1.4.13) leads directly to a bound on the total cross-section at high energy. The theorem, which is valid for all energies, takes the form

$$\sigma_{\text{total}} = \frac{\operatorname{Im} \tilde{F}(s, \cos\theta = 1)}{2q\sqrt{s}}, \qquad (1.4.15)$$

when the two particles in the initial state have equal mass. Here q is their centre-of-mass momentum. The theorem is a direct consequence of unitarity; the probability of a transition occurring out of an initial state $|n\rangle$ is

$$\sum_m |\langle m|R|n\rangle|^2, \qquad (1.4.16)$$

where R is the non-trivial part of S, as defined in (1.2.13). But, from the unitarity relation $S^\dagger S = 1$ and the completeness relation

$$\sum_m |m\rangle\langle m| = 1,$$

(1.4.16) is just $2\operatorname{Im}\langle n|R|n\rangle,$

which, by the definition (1.2.18) of F, is proportional to

$$\operatorname{Im} \tilde{F}(s, \cos\theta = 1).$$

The other factors in (1.4.15) are kinematical, arising from the definition of the cross-section.

Combining (1.4.15) with (1.4.13) we have, as $s \to \infty$,

$$\sigma_{\text{total}} < \text{constant} (\log s)^2.$$

1.5 Feynman diagrams and the S-matrix

We have remarked earlier that in the first three chapters we will be using the perturbation expansion of S-matrix elements, not for their numerical values but for their analytic properties. We believe that the singularities of these perturbation terms will in general be present on the physical sheet of the full S-matrix element. The situation on other sheets is more complicated and individual terms of the series may not give singularities that agree with those derived by unitarity. Sometimes also individual terms may give singularities on unphysical sheets that are cancelled when a sum of terms is taken (Landshoff, 1963).

Generally, individual terms in perturbation theory do not yield the correct singularities on unphysical sheets where some 'dynamical' property is involved, for example the formation of a resonance. However, on the physical sheet and on its boundary all singularities refer to stable particles and, whether these are poles or branch cuts, they do appear in individual terms of the perturbation series. The properties on unphysical sheets can be considered by S-matrix theory, but frequently they can be obtained more easily by a suitable choice of a summable series of matrix elements, for example by using the Bethe–Salpeter equation for a scattering amplitude.

For full details of the S-matrix expansion in terms of Feynman integrals the reader is referred to Schweber (1961) or to the original papers quoted there. Here we will give only the briefest indication of the method. The S-matrix can be expressed in operator form as

$$S = \sum_{n=0}^{\infty} \left(\frac{-i}{\hbar c} \right)^n \frac{1}{n!} \int_{-\infty}^{\infty} d^4x_1 \ldots d^4x_n T\{H_I(x_1) \ldots H_I(x_n)\}, \quad (1.5.1)$$

where T denotes a time-ordered product of the interaction Hamiltonian density $H_I(x)$ at the space-time points $x_1 \ldots x_n$. We will not specify the form of this Hamiltonian in general, but note just the example of a neutral pseudo-scalar coupling. For this example,

$$H_I(x) = GN\{\overline{\psi}(x) \gamma_5 \psi(x) \phi(x)\}, \quad (1.5.2)$$

where G is the coupling constant and N denotes a 'normal product' in which creation operators lie to the left of destruction operators. The fields ψ and ϕ are in the interaction representation, so they satisfy free-field commutation rules.

The matrix elements of S can be evaluated by rearranging the field operators in equation (1.5.1) into normal products by means of the commutation relations. Thus

$$T\{\phi(x_1) \phi(x_2)\} = N\{\phi(x_1) \phi(x_2)\} + \tfrac{1}{2}\hbar c \Delta_F(x_1 - x_2), \quad (1.5.3)$$

where $\Delta_F(x)$ is the Feynman propagator for the meson field,

$$\Delta_F(x) = \frac{2i}{(2\pi)^4} \int \frac{d^4k \exp{(-ikx)}}{k^2 - \mu^2 + i\epsilon}. \quad (1.5.4)$$

The rearrangement, of the expression (1.5.1) for S, into normal products permits the selection of only those terms whose creation and

destruction operators match the particles in the matrix elements that
we wish to consider. Then each term of (1.5.1) can be expressed as a
sum of terms involving propagators like (1.5.4), other propagators for
fermions, and factors arising for example from the γ-matrix in (1.5.2).
The space-time integrations can then be carried out, and they lead to
linear relations between the four-momenta that enter through such
expressions as (1.5.4).

The result is that the S-matrix element for a given scattering
process is the sum of a set of integrals over four-momenta. These are
the Feynman integrals with which our analytic studies are concerned.
They can be specified most easily by first writing down all Feynman
diagrams and then applying the Feynman rules, which we enumerate
below.

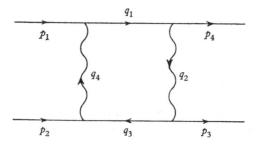

Fig. 1.5.1. An example of a Feynman diagram (or Feynman graph). The symbols
denote energy-momentum four-vectors, the straight lines denote nucleon lines and
the wavy lines denote meson lines.

An example of a Feynman diagram† is given in figure (1.5.1).
More generally an allowed Feynman diagram consists of any diagram
having the right number of external lines and with vertices that
correspond to the fields that occur in each term of the Hamiltonian.
Thus our example of an interaction in equation (1.5.2) has at each
vertex two fermion lines and one meson line. In Fig. 1.5.1 the straight
lines are fermion lines and the wavy lines are meson lines. The complete
S-matrix element corresponds to the sum of all topologically different
Feynman diagrams that have the correct number of external lines
and the type of vertex and internal lines specified by the assumed
interaction Hamiltonian.

Corresponding to a given Feynman diagram there is a contribution
to the S-matrix given by the following rules: It is an integral taken

† Note that we also use the term Feynman graph instead of Feynman diagram;
both are common in the literature of quantum field theory.

over the four-momentum of each internal line in the diagram, these four-momenta being labelled independently in the first instance.

The integrand contains the following factors:

(1) a factor $\dfrac{i}{(2\pi)^4}\dfrac{1}{k^2-\mu^2+i\epsilon}$ for each internal meson line of four-momentum k;

(2) a factor $\dfrac{i}{(2\pi)^4}\dfrac{\gamma p+m}{p^2-m^2+i\epsilon}$ for each internal nucleon line of four-momentum p;

(3) a factor $G\Gamma$ for each meson-nucleon vertex (where, for example, $\Gamma = \gamma_5$ for pseudo-scalar interaction as in (1.5.2));

(4) a factor 1 for each external meson line;

(5) a factor $\bar{u}_s(p)$ for each external nucleon line leaving the diagram, and a factor $u_s(p)$ for each external nucleon line entering the diagram (the Dirac spinor is normalised by $\bar{u}u = 2m$);

(6) a factor (-1) for each closed nucleon loop;

(7) a factor $(-i)^n$ corresponding to the diagram being nth order in the perturbation expansion;

(8) a factor $(2\pi)^4\delta^{(4)}(p-p'\pm k)$ for each vertex, so as to give energy-momentum conservation for a nucleon line p entering and p' leaving the vertex and a meson line k entering (or $-k$ leaving) the vertex.

The S-matrix element for a given process is the sum of all the Feynman integrals allowed for that process.

In practice because of the rule (8) given above many four-momenta can trivially be integrated. The remaining variables of integration can be regarded as circulating four-momenta following the method of Maxwell for taking account of Kirchoff's first law. In this case the total four-momentum entering each vertex must be conserved, as follows when we do the integration to eliminate the δ-functions in rule (8). One δ-function remains, expressing overall energy-momentum conservation; this can be factored out as in (1.2.18) so passing from the S-matrix element to the Feynman amplitude F.

As an example we take the diagram shown in Fig. 1.5.2, in which all the particles are scalar mesons and where we also assume a scalar interaction ($\Gamma = 1$). This gives the Feynman integral

$$-i\frac{G^4}{(2\pi)^4}\int d^4k \frac{1}{[(k+p_1)^2-m_1^2+i\epsilon]\,[(k-p_4+p_1)^2-m_2^2+i\epsilon]} \cdot$$
$$\times [(k-p_2)^2-m_3^2+i\epsilon]\,[k^2-m_4^2+i\epsilon]$$

$$(1.5.5)$$

In this diagram we have allowed the possibility of the internal scalar mesons having different masses m_i ($i = 1, ..., 4$). The external momenta must, for a scattering amplitude, satisfy the mass-shell condition and these external masses may take different values, thus $p_1^2 = M_1^2$ and so on.

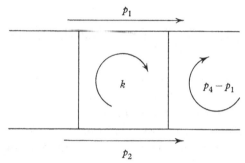

Fig. 1.5.2. The Feynman diagram that corresponds to the Feynman integral (1.5.5).

For an individual Feynman integral, the singularities, and hence the analytic structure of the integral, arise from singularities in the integrand. Those that arise from ultraviolet divergences are removed by renormalisation and will not concern us. The singularities that do concern us come from zeros of the denominator factors. These denominator factors are the same for scalar particles as they are for particles with spin, namely of the general form

$$p^2 - m_r^2 + i\epsilon \qquad (1.5.6)$$

for mass m_r and four-momentum p.

The numerator and the spinor factors given by the Feynman rules may lead to cancellation between different Feynman integrals, or within a single integral. However, these other factors are not expected to affect the possibility of a singularity in most of the terms. It is therefore advantageous to ignore all numerator terms and spin complications in the first instance. This permits the study of the analytic properties of individual Feynman integrals by the much more simple integrals that correspond to scalar particles, like the one in (1.5.5). If, however, we wish to add together Feynman integrals, for example by making a partial summation of the series, it is then essential to include the relevant numerator and spin factors in each integral. The latter point will be relevant in part of chapter 3.

For the rest of this section and for chapter 2 we will therefore work only with Feynman integrals that correspond to scalar bosons since, for each individual term, this gives the correct analytic properties. The general Feynman integral then takes the form

$$F = \lim_{\epsilon \to 0+} \int \frac{d^4k_1 \ldots d^4k_l}{\prod\limits_{r=1}^{n} (q_r^2 - m_r^2 + i\epsilon)}, \qquad (1.5.7)$$

where $k_1 \ldots k_l$ denote l independent loop four-momenta, and $q_1, \ldots, q_n$ denote the four-momenta for the n internal lines. Thus each q_s depends linearly on one or more of the k_j and possibly on one or more of the external four-momenta, which we will denote by p_a.

We remark in passing that for the simplest of all Feynman diagrams, that of Fig. 1.5.3, the energy-momentum conservation requirements

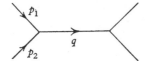

Fig. 1.5.3. A Feynman diagram leading to a pole in the series for a scattering amplitude.

leave no integrations to be done. Hence the diagram just corresponds to one of the poles in (1.5.7) with $q = p_1 + p_2$. The pole occurs at $s = q^2 = m^2$ on the physical sheet, and corresponds to the occurrence of a physical particle of mass m having the same quantum numbers as the system p_1 and p_2. If this term were taken alone, the pole would occur on every sheet in the variable s, since its structure is simple and is not affected by any branch-points or cuts through which we might pass to reach other sheets. However, it is easy to check from elastic unitarity that a pole cannot occur on the physical and the first un-physical sheet in the same position (see § 4.10). Therefore, on the unphysical sheet in the full S-matrix element this term must be cancelled by other terms. This is one example which prompted our cautionary remarks at the start of this section.

For computation (which does not concern us here) and for determination of analytic properties (which does concern us), it is often more convenient to transform the integral (1.5.7) by means of Feynman's identity,

$$\frac{1}{f_1 f_2 \cdots f_N} = (N-1)! \int_0^1 \frac{d\alpha_1 \ldots d\alpha_N \, \delta(\Sigma \alpha - 1)}{\left[\sum\limits_1^N \alpha_r f_r \right]^N}. \qquad (1.5.8)$$

This formula is valid provided that the denominator in the integral does not vanish within the range of α-integration (which otherwise would have to be distorted into the complex α_r-planes). If we take

$$f_r = q_r^2 - m_r^2 + i\epsilon, \qquad (1.5.9)$$

the denominator does not vanish because of the $i\epsilon$. Then the integral (1.5.7) is proportional to

$$\underset{\epsilon \to 0+}{\text{limit}} \int \frac{\left(\prod_1^l d^4 k_j \right) \left(\prod_1^n d\alpha_i \right) \delta(\Sigma\alpha - 1)}{\left[\sum_{r=1}^n \alpha_r(q_r^2 - m_r^2) + i\epsilon \right]^n}. \qquad (1.5.10)$$

It is now possible to do the k integrations (Chisholm, 1952). The four-momentum q_r in any internal line is a linear function of the circulating momenta k_j and the external momenta p_a. Therefore the quadratic form

$$\psi(k, p, \alpha) = \Sigma\alpha_r(q_r^2 - m_r^2) \qquad (1.5.11)$$

can be written as

$$\psi(k, p, \alpha) = \sum_{i,j=1}^l a_{ij} k_i k_j + \sum_{j=1}^l b_j k_j + c, \qquad (1.5.12)$$

$$= \mathbf{k}^T . \mathbf{Ak} - 2\mathbf{k}^T . \mathbf{Bp} + (\mathbf{p}^T . \mathbf{\Gamma p} - \sigma), \qquad (1.5.13)$$

where

$$\sigma = \Sigma\alpha_r m_r^2. \qquad (1.5.14)$$

Here, $\mathbf{A}$, $\mathbf{B}$, $\mathbf{\Gamma}$ are respectively $l \times l$, $l \times (E-1)$ and $(E-1) \times (E-1)$ matrices, whose elements are linear combinations of the α-variables.

E denotes the number of external lines of the diagram that gives the integral (1.5.10), so there are $(E-1)$ independent external four-momenta p_a occurring in ψ; $\mathbf{k}$ and $\mathbf{p}$ are column vectors in the spaces of the matrices, thus $\mathbf{k}$ has l components, namely the l four-vectors k_j and $\mathbf{p}$ has $(E-1)$ components, namely p_a; $\mathbf{k}^T$ and $\mathbf{p}^T$ denote the row vectors that are the transposes of $\mathbf{k}$ and $\mathbf{p}$. The elements of $\mathbf{k}$ are themselves Lorentz four-vectors, so the expressions in (1.5.13) correspond to double sums, thus

$$\mathbf{k}^T . \mathbf{Ak} = \sum_{\mu} \sum_{jj'} k_j^\mu A_{jj'} k_{\mu j'}, \qquad (1.5.15)$$

over both the matrix indices j, j' and the Lorentz index μ.

By a translation

$$\mathbf{k} = \mathbf{k}' + \mathbf{A}^{-1}\mathbf{Bp}, \qquad (1.5.16)$$

the terms of first degree in k in (1.5.13) can be eliminated. The resulting quadratic expression $\mathbf{k}'^T.\mathbf{A}\mathbf{k}'$ can be diagonalised by an orthogonal transformation in the matrix space,

$$\mathbf{k}' = \mathbf{R}\mathbf{k}'', \quad \mathbf{R}^T\mathbf{R} = 1, \tag{1.5.17}$$

$$\mathbf{R}^T\mathbf{A}\mathbf{R} = \mathbf{A}'' \text{ (diagonal)}, \tag{1.5.18}$$

giving $\quad \psi = \mathbf{k}''^T.\mathbf{A}''\mathbf{k}'' - (\mathbf{B}\mathbf{p})^T.\mathbf{A}^{-1}(\mathbf{B}\mathbf{p}) + \mathbf{p}^T.\mathbf{\Gamma}\mathbf{p} - \sigma.$ $\quad$ (1.5.19)

The Jacobian of the transformation $k \to k''$ is unity, that is

$$\int_{-\infty}^{\infty} \Pi d^4 k_j = \int_{-\infty}^{\infty} \Pi \, d^4 k_j''. \tag{1.5.20}$$

Having diagonalised the quadratic terms in the variables k_j'', we may now do all the energy-momentum integrations by repeated application of the formula

$$\int_{-\infty}^{\infty} \frac{du}{(Cu^2 + L)^s} \propto \frac{1}{C^{\frac{1}{2}} L^{s-\frac{1}{2}}}. \tag{1.5.21}$$

The result is proportional to

$$\lim_{\epsilon \to 0+} \int_0^1 \frac{\left(\prod_1^n d\alpha_r \right) \delta(\Sigma\alpha - 1) \, C^{n-2l-2}}{[D + i\epsilon C]^{n-2l}}, \tag{1.5.22}$$

where $\qquad\qquad\qquad C = \det\mathbf{A}'' = \det\mathbf{A},$ $\qquad\qquad$ (1.5.23)

$$D(p, \alpha) = -(\mathbf{B}\mathbf{p})^T.\mathbf{X}(\mathbf{B}\mathbf{p}) + (\mathbf{p}^T.\mathbf{\Gamma}\mathbf{p} - \sigma)\,C, \tag{1.5.24}$$

with $\qquad\qquad\qquad \mathbf{X} = \mathbf{A}^{-1}C = \operatorname{adj}\mathbf{A},$ $\qquad\qquad$ (1.5.25)

and σ defined by (1.5.14).

It can readily be seen that an alternative expression for D is

$$D = CD', \tag{1.5.26}$$

where D' is the result of eliminating k from ψ by means of the equations

$$\frac{\partial \psi}{\partial k_j} = 0 \quad \text{for each } j. \tag{1.5.27}$$

Note that for each j, this denotes four equations, one for each Lorentz component of k_j. These equations are just

$$\mathbf{A}\mathbf{k} = \mathbf{B}\mathbf{p}, \tag{1.5.28}$$

so they give $\qquad\qquad\qquad \mathbf{k} = \mathbf{A}^{-1}\mathbf{B}\mathbf{p}.$ $\qquad\qquad$ (1.5.29)

Insertion of this into (1.5.13) gives, with (1.5.26), the result (1.5.24).

The discriminant $D(p, \alpha)$ is of great importance in the study of analytic properties of Feynman integrals. In practice there are two ways to obtain its detailed form for any given diagram. The first way is to proceed by labelling all the lines with their appropriate four-momenta, taking account of conservation at each vertex in the diagram as in Fig. 1.5.2.

This permits explicit evaluation from (1.5.11) of the coefficients a_{ij} in equation (1.5.12) which are linear functions of the α_r; and of the coefficients b_j which are linear in the α_r and linear in the external momenta p; and of c which is also linear in the α and quadratic in the external momenta and the internal masses. With explicit forms for these coefficients C and D are given by

$$C(\alpha) = \begin{vmatrix} a_{11} & \cdots & a_{1l} \\ \vdots & \vdots & \vdots \\ a_{l1} & \cdots & a_{ll} \end{vmatrix} = \det \mathbf{A}, \qquad (1.5.30)$$

$$D(p,\alpha) = \begin{vmatrix} a_{11} & \cdots & a_{1l} & b_1 \\ \vdots & \vdots & \vdots & \vdots \\ a_{l1} & \cdots & a_{ll} & b_l \\ b_1 & \cdots & b_l & c \end{vmatrix}. \qquad (1.5.31)$$

It will be noticed that D is a function of α and of the invariants formed by the scalar products of the external four-momenta

$$p_a \, (a = 1, ..., (E-1))$$

C is of degree l in the α and D of degree $(l+1)$.

The second method for obtaining $C(\alpha)$ and $D(p,\alpha)$ will be presented as a set of rules, which permit these functions to be written down directly for any given Feynman diagram. We state these rules without proof; their derivation is similar to that used by Symanzik (1958) in a slightly different formalism.

These rules are most simply described by means of a particular example, for which we use the diagram in Fig. 1.5.4a. In this figure, the numbers on the internal lines correspond to the indices in the above equations.

We first give the rules for writing down the function $C(\alpha)$. A set of cuts is made in the internal lines, such that: (i) every vertex is still connected to every other vertex by a sequence of uncut lines; (ii) no further cuts can be made without violating (i). An example of a possible set of cuts is shown in Fig. 1.5.4b, where the cut lines are marked with crosses. The set of cuts corresponds to a contribution to C that is the product of the α_i corresponding to the cut lines. Thus the cuts in Fig. 1.5.4b yield for C a contribution $\alpha_1\alpha_6$. The complete expression for C arises from adding in the products of α for every possible set of cuts that obey (i) and (ii). Hence in our example

$$C = \alpha_1(\alpha_2+\alpha_4+\alpha_6+\alpha_7)+\alpha_2(\alpha_3+\alpha_5+\alpha_6)+\alpha_3(\alpha_4+\alpha_6+\alpha_7)$$
$$+\alpha_4(\alpha_5+\alpha_6)+\alpha_5(\alpha_6+\alpha_7)+\alpha_6\alpha_7. \quad (1.5.32)$$

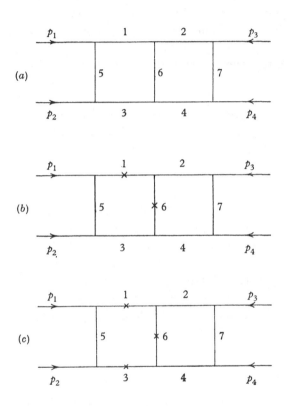

Fig. 1.5.4. Examples of possible sets of cuts (denoted by a cross on the cut line) used in evaluating the integrand using Feynman parameters, (a) the diagram, (b) cuts for evaluating a term in C, (c) cuts for evaluating a term in D.

We now give the rules of writing down the remainder of the expression (1.5.24), that is

$$-(\mathbf{Bp})^T.\mathbf{X}(\mathbf{Bp})+C\mathbf{p}^T.\mathbf{\Gamma p}. \quad (1.5.33)$$

This expression is a linear function of the various Lorentz scalars that can be manufactured from the external momenta p.

The rules are: By a set of cuts on the internal lines divide the graph into two disjoint parts, such that within each part the previous rules

(i) and (ii) are obeyed and such that at least one external momentum-line is connected to each part. Any such set of cuts yields a product of α_i for the two disjoint parts, exactly as before. These products are multiplied together and are also multiplied by the Lorentz scalar $(\Sigma'p)^2$, where $(\Sigma'p)$ denotes the sum of the external momenta coming into either of the disjoint parts (because of energy-momentum conservation, $(\Sigma'p)^2$ is the same for both parts). Do not forget the possibility that $(\Sigma'p)$ may include only the four-momentum in one external line for an appropriate set of cuts.

As an example, the set of cuts drawn in Fig. 1.5.4c yields the contribution

$$\alpha_1 \alpha_3 \alpha_6 (p_1 + p_2)^2.$$

The complete expression for the diagram in Fig. 1.5.4a is

$$
\begin{aligned}
(p_1 + p_2)^2 & \left[\alpha_1 \alpha_3 (\alpha_2 + \alpha_4 + \alpha_6 + \alpha_7) + \alpha_2 \alpha_4 (\alpha_1 + \alpha_3 + \alpha_5 + \alpha_6) \right. \\
& \left. + \alpha_2 \alpha_3 \alpha_6 + \alpha_1 \alpha_4 \alpha_6 \right] + (p_1 + p_3)^2 \left[\alpha_5 \alpha_6 \alpha_7 \right] \\
& + p_1^2 \left[\alpha_1 \alpha_5 (\alpha_2 + \alpha_4 + \alpha_6 + \alpha_7) + \alpha_2 \alpha_5 \alpha_6 \right] \\
& + p_2^2 \left[\alpha_3 \alpha_5 (\alpha_2 + \alpha_4 + \alpha_6 + \alpha_7) + \alpha_4 \alpha_5 \alpha_6 \right] \\
& + p_3^2 \left[\alpha_2 \alpha_7 (\alpha_1 + \alpha_3 + \alpha_5 + \alpha_6) + \alpha_1 \alpha_6 \alpha_7 \right] \\
& + p_4^2 \left[\alpha_4 \alpha_7 (\alpha_1 + \alpha_3 + \alpha_5 + \alpha_6) + \alpha_3 \alpha_6 \alpha_7 \right] \quad (1.5.34)
\end{aligned}
$$

These general rules are of particular value when one needs, as in chapter 3, the coefficient of a particular external variable, sometimes for quite complicated diagrams where the whole expression for D would be immense. For simpler diagrams when one needs the full expression for D the form (1.5.31) obtained by direct calculation is often more convenient to use.

1.6 Applications

We conclude this chapter by giving a brief indication of some of the applications of S-matrix theory to the physics of particles.

The simplest and best known use of analyticity of a scattering amplitude is given by the Breit–Wigner resonance formula. Near a resonance an amplitude may be approximated by neglecting all singularities except a complex pole. Thus the partial amplitude for an S-wave may have the form

$$A(k) = \frac{2ib}{k - a + ib}, \quad (1.6.1)$$

where k is the numerical value of the three-momentum. This non-relativistic formula gives a simple illustration of the approximation of a transition amplitude by including nearby singularities and ignoring distant singularities. This kind of approximation provides the basis for most successful applications of S-matrix theory.

Another well-known result that follows directly from a study of the nearby singularities of the S-matrix is the scattering-length and effective-range formula. We shall see in §4.7 that the singularity of an elastic scattering amplitude at threshold (zero energy) is a simple square-root branch-point in energy. Thus there will be no branch cut in the momentum variable k. If we assume a nearby bound state (or virtual state) the non-relativistic s-wave, for example, takes the form

$$e^{2i\delta} = -\left(\frac{k+ib}{k-ib}\right)(1+ak+\ldots),\qquad (1.6.2)$$

where we assume the momentum is small and that there are no other nearby singularities. This gives the usual formula

$$k\cot\delta = c_0 + c_1 k^2.\qquad (1.6.3)$$

An application of analyticity methods to the electromagnetic structure of nucleons by Frazer & Fulco (1959, 1960a, b) provided a great stimulus both to the development of S-matrix theory and to experimental work. For electron-proton scattering one works to first order in the fine structure constant. Then from Lorentz invariance the scattering amplitude can depend only on the square of the four-momentum of the exchanged photon shown in Fig. 1.6.1a

$$t = q^2 = q_0^2 - \mathbf{q}^2.\qquad (1.6.4)$$

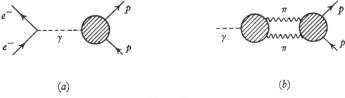

(a)   (b)

Fig. 1.6.1.

Neglecting spin, the amplitude is

$$A(t) = \frac{e^2}{t}\{1 + F(t)\}\qquad (1.6.5)$$

The term $F(t)$ represents the modification to the Coulomb interaction due to the structure of the photon-nucleon vertex indicated in

Fig. 1.6.1 a. Since this vertex involves strong interactions the methods of dispersion theory must be used and one looks for the nearest singularities in the complex t-plane. These include the branch cut that arises from the creation of a pion pair as indicated in Fig. 1.6.1 b. This branch cut runs from $4m^2$ to ∞ in the t-plane and may be contrasted with the physical values of t for electron-proton scattering which are real and negative.

It can be shown (§ 2.5) that $F(t)$ to all orders in perturbation theory is analytic in the complex t-plane cut along the two-pion branch cut. Assuming suitable convergence at infinity this gives a dispersion relation for $F(t)$, namely,

$$F(t) = \frac{1}{\pi} \int_{4m^2}^{\infty} \frac{F_t(x)\,dx}{x-t}. \tag{1.6.6}$$

The spectral function $F_t(x)$ is directly related to the amplitude for pion-pion scattering (amongst other terms). By approximating this amplitude with a resonance Frazer and Fulco were able to obtain good agreement with experimental results for electron-nucleon scattering.

More recently these methods for approximating the analytic functions of S-matrix theory have been refined and applied to most experimental situations involving the pion-nucleon system and also to other strongly interacting particles. For a review of applications to the pion-nucleon system see Hamilton (1964), and for a general account of the applications of the S-matrix see Frautschi (1963), and Omnes & Froissart (1964).

There is a possibility that the analytic properties of the S-matrix can be used in a more fundamental way to set up a self-consistency condition when combined with unitarity and symmetry properties. If it could be solved in a satisfactory manner such a self-consistency condition could yield relations between the masses of strongly interacting particles including the position and width of resonances. There are some very difficult mathematical problems associated with the rigorous justification of approximations associated with such a self-consistent scheme and we do not consider it further in this book though we note its potential importance. A discussion of these aspects of application of S-matrix theory is given by Chew (1962) and Chew & Jacob (1964).

CHAPTER 2

ANALYTIC PROPERTIES OF PERTURBATION THEORY

2.1 Singularities of integral representations

We have said in chapter 1 that a transition amplitude may be expanded in a perturbation series, each term in the series being a Feynman integral. Since the perturbation series cannot necessarily be expected to converge, we attach no significance to the numerical values of the individual terms, but rather examine their analytic properties. In this chapter we develop general methods for doing this. We illustrate these methods by applying them to individual Feynman diagrams and even, in some instances, to every diagram occurring in the perturbation series for a given amplitude.

In general a Feynman integral is too complicated for its analytic properties to be studied by explicit integration. For example, one of the very simplest diagrams, the fourth-order single loop of Fig. 1.5.2 with all scalar particles, produces 192 Spence functions (see the paper by A. C. T. Wu (1961)). But fortunately an easier and more powerful method is available. It is based on a generalisation of a lemma due to Hadamard (1898), which was re-discovered in the present context by Eden (1952) and extended by Polkinghorne & Screaton (1960a,b). This lemma is concerned with the circumstances in which a function defined by an integral may have a singularity. In the remainder of this section we develop the main ideas behind the method, first for the case of one integration variable and then for several.

Singularities of simple integrals

Let $g(z, w)$ be an analytic function of two complex variables and let C be some finite contour in the complex w-plane. Define a function $f(z)$ by

$$f(z) = \int_C g(z, w) \, dw. \qquad (2.1.1)$$

It is supposed that the singularities of the integrand g are known and that their locations in the w-plane are

$$w = w_r(z) \quad (r = 1, 2, \ldots). \qquad (2.1.2)$$

It is further supposed that for z in a neighbourhood of some point z_0 there is a neighbourhood of the contour C in the w-plane free from the singularities w_r. Then evidently the definition (2.1.1) makes $f(z)$ analytic at z_0. We wish to discover how a singularity of $f(z)$ can arise when f is analytically continued away from z_0.

As z is moved away from z_0 the singularities w_r will move about in the w-plane, but $f(z)$ will remain analytic at least until one of the w_r reaches the contour C. Then the integral (2.1.1) becomes undefined. But even when this has happened, $f(z)$ can usually be further analytically continued. Suppose that C has end-points A and B. We suppose at first that these end-points are fixed. Then if C' is another contour with the same end-points and if both C' and the region between C and C' are free of the singularities w_r, Cauchy's theorem gives

$$f(z) = \int_{C'} g(z, w)\, dw. \qquad (2.1.3)$$

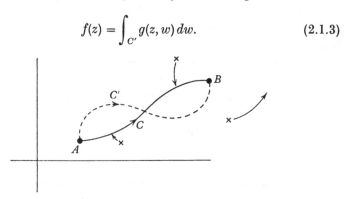

Fig. 2.1.1. The complex w-plane. Analytic continuation by deforming the contour of integration C to C'. Singularities of the integrand are denoted by $\times$, and arrows indicate how they move as z in equation (2.1.3) is varied.

So just before one of the w_r reaches the contour C we can usually change to another contour C', and equation (2.1.3) then provides an analytic continuation of the function originally defined by (2.1.1). This is illustrated in Fig. 2.1.1; here the crosses represent typical positions of the various singularities w_r at the original value z_0 of z and the arrows indicate how they might have moved during the variation of z.

So generally we can continue $f(z)$ analytically if, by varying the integration contour C continuously, we can arrange that as z is varied, none of the singularities $w_r(z)$ of the integrand $g(z, w)$ meets the contour. This procedure may be prevented for one of three reasons:

(i) *End-point singularities.* If one of the singularities w_r reaches one of the end-points A, B of the contour C no allowed variation of C can avoid it. Hence the corresponding point z_1 may be a singularity of $f(z)$.

(ii) *Pinch singularities.* If two (or more) singularities approach the contour from opposite sides and coincide, the contour C will be trapped between them and no deformation can avoid them.† Hence the corresponding point z_2 may be a singularity of $f(z)$.

(iii) *Infinite deformations.* If the contour C is being deformed to avoid a singularity $w_r(z)$, and that singularity moves off to infinity dragging C with it, the integral may diverge because C is no longer finite. This can be reduced to a special case of (ii); it is best examined by making a transformation of integration variable, for example $w = 1/\zeta$, so that the point at infinity is brought into the finite part of the complex plane.

Examples of the above possibilities are given by the following:

(i)
$$f(z) = \int_a^b \frac{dw}{w-z} = \log\left(\frac{b-z}{a-z}\right). \qquad (2.1.4)$$

Here the integrand $(w-z)^{-1}$ is singular at the end-points a, b if $z = a, b$, corresponding to the singularity of the logarithm at these points.

(ii) $\quad f(z) = \int_0^1 \frac{dw}{(w-z)(w-a)} = \frac{1}{z-a}\log\left\{\frac{a(1-z)}{(1-a)z}\right\} \quad (a>1). \quad (2.1.5)$

Here the logarithmic singularities $z = 1, 0$ are end-point singularities, while the singularity $z = a$ arises from a pinch between the singularities $w_1 = z$, $w_2 = a$ of the integrand. Notice, however, that the singularity $z = a$ is not present on the Riemann sheet of $f(z)$ corresponding to the principal sheet of the logarithm, for the argument of the logarithm is unity at $z = a$. The singularity is only encountered on encircling one of the logarithmic singularities, so that now $\log 1$ is $\pm 2\pi i$ instead of zero and no longer cancels the pole. This example underlines the importance of pinch singularities actually trapping the contour by approaching it from opposite sides, as in Fig. 2.1.2a; if they approach from the same side or if they come together nowhere near the contour, as in Fig. 2.1.2b they are harmless. Finding whether the contour is actually trapped is in practice the difficult part of the analysis.

---

† Alternatively, one of the two singularities w_r may be fixed and the trapping may occur as a result of the other approaching it, with C between.

(iii)
$$f(z) = \int_2^3 \frac{dw}{zw+1} = \frac{1}{z}\log\left(\frac{3z+1}{2z+1}\right).$$
(2.1.6)

The singularities $z = -(\frac{1}{2})$, $-(\frac{1}{3})$ are end-point singularities. On the principal Riemann sheet of the logarithm there is no singularity $z = 0$, but on other sheets it arises as a result of the contour in the w-integration being dragged to infinity. To see this, make the transformation of integration variable $w = 1/\zeta$:

$$f(z) = \int_{\frac{1}{3}}^{\frac{1}{2}} \frac{d\zeta}{\zeta(\zeta+z)}.$$
(2.1.7)

The singularity $z = 0$ now arises from a pinch at $\zeta = 0$.

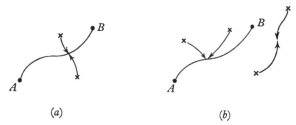

(a)　　　　　　　　　　　　(b)

Fig. 2.1.2. (a) Two singularities in the w-plane coming together and pinching the contour of integration. (b) Examples of coincidence of two singularities that do *not* pinch the contour of integration.

We supposed in the above that the end-points A, B of the contour C were fixed, but there is in principle no new feature if one or both is allowed to be a function of z. Pinch singularities may occur as before, while an end-point singularity can occur at $z = z_1$, if for some r,

$$\left.\begin{aligned} w_r(z_1) &= A(z_1), \\ w_r(z_1) &= B(z_1). \end{aligned}\right\}$$
(2.1.8)

or if

Again, the contour C may be closed and have no end-points at all. In this case only pinch singularities are encountered. Finally, we may drop the restriction that C be initially finite, because of our remark that by a transformation of variables we can see that the point at infinity is no different from any other point.

Nature of the singularities

The singularities produced by the above mechanisms are usually branch-points. It is convenient to attach a cut to a branch-point and in certain simple cases one can easily determine the discontinuity of $f(z)$ across a given cut, in particular when the singularities w_r of the integrand $g(z, w)$ are just poles. We suppose, as an illustration, that A is fixed and that $z = z_1$ is an end-point singularity produced by w_1:

$$w_1(z_1) = A.$$

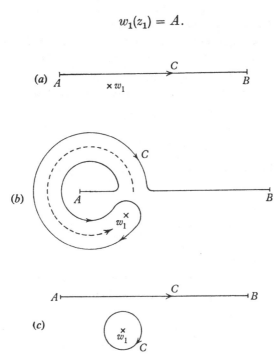

Fig. 2.1.3. The movement of a singularity $w_1(z)$ of the integrand of equation (2.1.3) when the variable z encircles an end-point singularity z_1. The starting position is shown in (a), the end-position in (b) and (c) which are equivalent.

Then, when z is at some point $z_1 + \epsilon$ very near to z_1, $w_1(z)$ will be very near to A. If $w_1(z)$ is analytic at z_1, and has non-zero derivative there, we can write

$$w_1(z_1 + \epsilon) = A + \epsilon \frac{dw_1}{dz}\bigg|_{z=z_1} \qquad (2.1.9)$$

approximately. Thus if z is made to describe a complete circle round z_1: $z = z_1 + |\epsilon| e^{i\theta}$, θ varying from 0 to 2π, w_1 will describe a similar circle, of radius $|\epsilon\, dw_1/dz|$, round A. The contour C must be distorted to avoid w_1, as it describes its circle, so that if the situation in the w-plane before the circle is described is as in Fig. 2.1.3a, afterwards it will be as in Fig. 2.1.3b. But, by Cauchy's theorem, this is equivalent to Fig. 2.1.3c. Hence the difference between the initial and final integrals is just a contour round the pole $w_1(z)$. So the discontinuity of $f(z)$ associated with the singularity at $z = z_1$, is just $2\pi i$ times the residue of $g(z, w)$ at the pole $w = w_1$. Further, the same discontinuity is picked up each time z_1 is encircled, indicating that the $z = z_1$ singularity of $f(z)$ is logarithmic.

Tracing how the singularities w_r move as z is varied also enables us to understand how a pinch singularity may be absent on one Riemann sheet associated with an end-point singularity, but be present on others. This was the case in our example (2.1.5). Let $z = z_2$ be the singularity of $f(z)$ arising from a pinch between w_1 and w_2:

$$w_1(z_2) = w_2(z_2),$$

and let $z = z_1$ be the end-point singularity arising from

$$w_1(z_1) = A,$$

where again A is fixed. Then near $z = z_2$ the situation in the w-plane will be as in Fig. 2.1.4a. If now z is made to encircle z_1 and return to near z_2, w_1 will in general move round A and so the situation will be as in Fig. 2.1.4b. Hence the pinch no longer traps the contour and the pinch singularity is not present on this Riemann sheet.

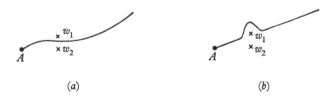

(a)   (b)

Fig. 2.1.4. A change in the relative position of a singularity $w_1(z)$ that corresponds to a change in z from one Riemann sheet to another. In the first position (a) there is a pinch of the contour; in the second (b) there is no pinch and no consequent singularity in the z-plane.

It should be stressed that the arguments of the last two paragraphs assume that the very simplest possibilities occur, and in practice each individual integral has to be examined in detail. Thus, if $dw_1/dz = 0$ in (2.1.9), though not d^2w/dz^2, w_1 will go twice round A when z makes a complete circuit round z_1. Or in the analysis of the pinch singularity, other w_r may get in the way while z is encircling z_1. And in either case if A varies with z things may be very different.

More than one external variable

The previous discussion generalises very easily to a function of more than one complex variable defined by a simple integral. As an example, consider

$$f(z, z') = \int_C g(z, z', w)\, dw. \qquad (2.1.10)$$

Except that the positions of the integrand's singularities w_r are now generally functions of two variables z, z', the conditions for a singularity of f are just as before. Thus, end point singularities are given by

$$w_r(z, z') = A \qquad (2.1.11)$$

for some r, and w_1 and w_2 can produce a pinch singularity when

$$w_1(z, z') = w_2(z, z'). \qquad (2.1.12)$$

Each of (2.1.11) and (2.1.12) defines a two-dimensional† surface Σ in the four-dimensional space of the complex variables z, z'.

(a)   (b)

Fig. 2.1.5. An example (a) of how a pinch singularity of equation (2.1.2) falls off the contour of integration C for one path in the (z, z')-space, whilst for another path the pinch may be retained as in (b).

Suppose now that (z_0, z_0') is some point on the surface Σ that is singular by virtue of (2.1.12), that is at (z_0, z_0') the contour C is pinched by w_1 and w_2. Suppose now we move to some neighbouring point on Σ. Then (2.1.12) is still satisfied, though the pinch in general now occurs at a different point in the w-plane. That is, the pinch has moved, dragging the contour C with it if necessary. This situation will persist if we move around anywhere on the surface Σ that is the solution to (2.1.12) provided neither one of two things happens. The simpler of the two things is that the pinch falls off the end of the contour C, as illustrated in Fig. 2.1.5a. But when w_1 and w_2 are at A we have just the condition for an end-point singularity. Hence it seems that at the intersection of the pinch surface Σ with an end-point singularity

† Because each represents two equations, one involving Re w and one Im w and so yields two constraints on the four variables Re z, Im z, Re z', Im z'.

surface Σ_1 the former surface can change from being singular to non-singular. In other words, it seems that the surface Σ that corresponds to the pinch equations (2.1.12) can be divided into singular and non-singular parts by its intersection with the end-point singularity surface (or surfaces) Σ_1; on the non-singular parts the pinch still occurs but the contour C is not trapped by it, as in Fig. 2.1.2b. Notice, however, that since Σ and Σ_1 are both two-dimensional surfaces in the four-dimensional complex (z, z') space, their intersections are just a set of points. These are not sufficient to divide Σ into different regions, but to each of the points is attached a cut, since Σ_1 is usually a branch-point singularity. So it is really these cuts that divide Σ into singular and non-singular parts. The cuts, however, are arbitrarily positioned—only their ends (namely the singularity surface Σ_1) are defined—and so there is no *unique* division of Σ into singular and non-singular parts. The division depends essentially on the path of analytic continuation used.

This last point is illustrated by the observation that, although a pinch *may* fall off the end of C when its end-point is reached, it need not. In order to see unambiguously what happens to $f(z, z')$ at a branch-point one must not pass through the branch-point itself, but take a small detour round it. Thus in Fig. 2.1.5a we must suppose that when we get near the intersection of Σ with Σ_1, we must temporarily leave and take a detour round the intersection before returning to Σ. Then the pinch will separate a little and one can see which side of the end-point A, w_1 and w_2 pass. The side to which they pass will generally depend on the sense of the detour; if one makes a detour one way round the intersection of Σ and Σ_1 the result might be Fig. 2.1.5a, while for the other way it might be Fig. 2.1.5b. The latter situation illustrates that by going a different way round (2.1.12), the end-point singularity branch-point, the pinch is not lost after all.

Falling off the end of the contour is not, however, the only way in which a pinch may become harmless. Suppose that the w_r are the roots of some algebraic equation

$$S(w; z, z') = 0. \qquad (2.1.13)$$

Then the pinch condition (2.1.12) is just the condition for two roots of S to coincide. Hence it is equivalent to (2.1.13) together with

$$\frac{\partial S}{\partial w} = 0. \qquad (2.1.14)$$

Now generally there will be further roots of S, other than the pair involved in the pinch. Suppose, as we move around on Σ, we come to a point where a third root joins the pinch, as in Fig. 2.1.6a. At this point, in addition to (2.1.13) and (2.1.14),

$$\frac{\partial^2 S}{\partial w^2} = 0. \tag{2.1.15}$$

If we move away from this point where S has three coincident roots we are back with only a coincident pair, but it is not necessarily the case that we have the same pair as originally. One of the roots that originally formed the pinch may have left, leaving the other together with the newcomer, as in Fig. 2.1.6b. In this way the pinch can become

$$(a) \qquad\qquad (b)$$

Fig. 2.1.6. An example of how a pinch singularity may be lost by going through a multiple pinch situation. For the initial positions shown in (a) there is a pinch, but in the final position (b) there is no pinch.

harmless. To see whether this actually does happen in a given continuation one must again avoid the critical point in (z, z') space by a small detour. Again the nature and sense of the detour will be important. This mechanism will be discussed further in § 2.8, when it is found to arise in the particular Feynman integral which we consider in that section.

Singularities of multiple integrals

The previous discussion may be generalised to integral transforms that involve several variables of integration. Consider

$$f(z) = \int_H \Pi \, dw_i g(z, w_i). \tag{2.1.16}$$

Here the contour of integration C of the simple integral case has become a 'hypercontour' H in the multi-dimensional complex w_i-space. In most problems one will have the function f defined such that for some part R of the real axis in the z-plane the integration region H is some part of the real w_i-space, this being possible because no singularities of the integrand g lie in this part of the real w_i-space when $z \in R$. The objective is then to continue $f(z)$ away from R.

The singularities of the integrand $g(z, w_i)$ are imagined as being given by various equations

$$S_r(z, w_i) = 0 \quad (r = 1, 2 \ldots).$$

For any value of z a given S_r will be a $(2n-2)$-dimensional surface in the $2n$-dimensional complex w_i-space. When z is varied away from R these S_r will move around till one of them comes to intersect H. To avoid this, H must be distorted away from its original real location, much as the simple contour C was distorted for the case of the single integration variable. When the possibility of this distortion ceases, so that H becomes trapped, we are liable to have encountered a singularity of $f(z)$.

We list the cases when this happens. A proper proof needs the use of topology; the difficulty lies in imagining what happens in the four (or more)-dimensional space of the integration variables. We shall be content with plausibility arguments.

(a) Usually, when a singularity surface S_r advances on H a distortion of H away from S_r in the direction of the normal to S_r will keep H out of the way and so will avoid the singularity. But if two singularity surfaces S_1 and S_2 say, advance on H from opposite sides and the directions of their normals coincide, H may be trapped. The conditions for this are

$$S_1 = S_2 = 0, \tag{2.1.17a}$$

$$\alpha_1 \frac{\partial S_1}{\partial w_i} + \alpha_2 \frac{\partial S_2}{\partial w_i} = 0 \quad (i = 1, \ldots, n), \tag{2.1.17b}$$

for some α_1, α_2.

(b) Two different parts of the same singularity surface, S_1 say, may trap H. One may think of S_1 becoming locally cone-like, with H trapped at the vertex of the cone. The conditions for this are

$$S_1 = 0 = \frac{\partial S_1}{\partial w_i} \quad (i = 1, \ldots, n), \tag{2.1.18}$$

(c) More than two of the S_r may participate. The conditions for this are that the surfaces intersect at a point, and there is a linear relation among the directions of their normals:

$$S_1 = S_2 = S_3 = 0, \tag{2.1.19a}$$

and
$$\alpha_1 \frac{\partial S_1}{\partial w_i} + \alpha_2 \frac{\partial S_2}{\partial w_i} + \alpha_3 \frac{\partial S_3}{\partial w_i} = 0 \quad (i = 1, ..., n). \tag{2.1.19b}$$

With n integration variables at most $(n+1)$ of the S_r can participate. With $(n+1)$ surfaces (2.1.19b) is trivially satisfied.

One way to see that the conditions (2.1.19) are plausible is to consider the comparison function

$$F(z) = \int_H \Pi dw_i \frac{1}{S_1 S_2 S_3}. \tag{2.1.20}$$

The integrand has singularity surfaces coinciding with those of interest for g, and so these should produce the same distortions of H in both (2.1.20) and (2.1.16). But (2.1.20) may be transformed by Feynman's identity (1.5.8):

$$F(z) = 2 \int \frac{d\alpha_1 d\alpha_2 d\alpha_3 \, \Pi dw_i \, \delta(\Sigma\alpha - 1)}{\left(\sum\limits_{r=1}^{3} \alpha_r S_r\right)^3}. \tag{2.1.21}$$

Here the integrand contains the single singularity surface

$$D = \sum_{r=1}^{3} \alpha_r S_r = 0$$

in the space of the $n+3$ integration variables α_1, α_2, α_3, w_i. So the conditions for a singularity of F are, by analogy with (2.1.18),

$$D = 0 = \frac{\partial D}{\partial \alpha_r} = \frac{\partial D}{\partial w_i},$$

which are just the equations (2.1.19).

(d) Just as the distortions of the contour C in the simple integral case were not allowed to move the end-points of C, so here there are restrictions in the distortions allowed on the boundary of the hypercontour H. If we suppose that the boundary of H is specified by various analytic equations

$$\tilde{S}_r(z, w, w') = 0,$$

then only distortions within these manifolds are permitted on the boundary. Since the normals to the $\tilde{S}_r$ represent directions in which the hypercontour cannot move these surfaces play their role in trapping the hypercontour in a similar way to the S_r. Thus we treat S_r and $\tilde{S}_r$ surfaces on an equal footing in determining the singularity equations.

To sum up, all the possibilities can be expressed together by introducing numbers α_i, $\tilde{\alpha}_r$ and solving the equations

$$\alpha_i S_i = 0, \quad \text{for each } i, \tag{2.1.22a}$$

(so that either α_i or $S_i = 0$),

$$\tilde{\alpha}_r \tilde{S}_r = 0, \quad \text{for each } r, \tag{2.1.22b}$$

and, for each integration variable w_j,

$$\frac{\partial}{\partial w_j}\{\sum_i \alpha_i S_i + \sum_r \tilde{\alpha}_r \tilde{S}_r\} = 0. \tag{2.1.22c}$$

We reiterate that our discussion of multiple integrals is devoid of mathematical rigour. We have merely aimed to give the reader enough information to make plausible the main applications. A rigorous treatment requires homology theory and for this we refer to the paper by Fotiadi, Froissart, Lascoux & Pham (1964).

2.2   The Landau equations

When the general methods of §2.1 are applied to a Feynman integral a set of equations is obtained which, in principle, determines the location of the singularities in the complex space of the external momentum variables p. There are two main forms of these equations, depending on which representation of the Feynman integral is studied. Both forms of these sets of equations were given by Landau (1959a, b) at the 1959 Kiev Conference and they are usually called the Landau equations. These equations were also derived by Bjorken (1959), Mathews (1959) and Nakanishi (1959). These authors used methods having less general applicability than those based on the techniques described in §2.1. These techniques were used by Eden (1952) to discuss normal thresholds and in their extended form were used by Polkinghorne & Screaton (1960a, b) to derive the Landau equations.

We recall from §1.5 three different representations for a Feynman integral, in the case when all the N internal lines of the corresponding

Feynman graph represent spinless particles. Apart from constant multiplicative factors these are

$$I = \int \frac{d^4k_1 d^4k_2 \dots d^4k_l}{\prod\limits_{i=1}^{N}(q_i^2 - m_i^2)}, \qquad (2.2.1)$$

as in (1.5.7), and

$$I = \int \frac{d^4k_1 \dots d^4k_l \left(\prod\limits_{i=1}^{N} d\alpha_i\right) \delta(\Sigma\alpha_i - 1)}{[\psi(p, k, \alpha)]^N} \qquad (2.2.2)$$

as in (1.5.10), with ψ defined in (1.5.11) by

$$\psi(p, k, \alpha) = \sum_{i=1}^{N} \alpha_i(q_i^2 - m_i^2), \qquad (2.2.3)$$

and lastly the result of integrating over the l independent loop momenta k

$$I = \int_0^1 \frac{\left(\prod\limits_{i=1}^{N} d\alpha_i\right)\delta(\Sigma\alpha_i - 1)\,C^{N-2l-2}}{[D]^{N-2l}}. \qquad (2.2.4)$$

The rules for writing down C and D for a given graph we described at the end of § 1.5. C is just a sum of certain products of the α, while D is of the form

$$D = \sum_r f_r(\alpha)\,z_r - C(\alpha)\sum_i \alpha_i m_i^2. \qquad (2.2.5)$$

Here the z_r are the various scalar products that can be formed from the external momenta p of the graph (including the squares of the masses).

The integral I is a function of the z_r and for physical external momenta each z_r is restricted to a certain part of the real axis. To obtain the physical values of I we recall that each integral (2.2.1), (2.2.2) or (2.2.4) should be evaluated with real hypercontour of integration and a small negative imaginary part added to the square of each internal mass m_i. This guarantees that none of the denominators in the three representations for I vanishes when the z_r are physical and the hypercontours of integration are real and undistorted.

The task of this chapter is to investigate the analyticity properties of I when the z_r are regarded as complex variables. So we start with the z_r real and physical and the hypercontour undistorted, and then vary the z_r into the complex space, distorting the hypercontour as necessary to avoid singularities of the integrand. The conditions for the possibility of this distortion to cease were given in § 2.1, and we now apply them to each of the representations (2.2.1), (2.2.2) and (2.2.4) in turn.

First representation

Each k-integration is infinite, so the hypercontour in (2.2.1) has no boundary. Hence in (2.1.22) there are no $\tilde{S}_r$, while the S_r are just the factors $(q_i^2 - m_i^2)$. So we introduce a parameter α_i corresponding to each of these factors and (2.1.22) gives

$$either \quad q_i^2 = m_i^2, \tag{2.2.6a}$$

$$or \quad \alpha_i = 0, \tag{2.2.6b}$$

for each i, together with

$$\frac{\partial}{\partial k_j} \Sigma \alpha_i (q_i^2 - m_i^2) = 0, \tag{2.2.7a}$$

for each loop momentum integration variable k_j. Since each q is a linear combination of the k this is just

$$\sum_j \alpha_i q_i = 0 \quad \text{for each } j, \tag{2.2.7b}$$

where $\sum_j$ denotes summation round the loop around which k_j runs. The equations (2.2.6) and (2.2.7) are the Landau equations corresponding to the representation (2.2.1).

Second representation

In the representation (2.2.2) the integrand has a single surface S of singularity, namely $\psi = 0$, but the hypercontour now drawn in complex (k, α) space has various boundaries $\tilde{S}$, namely $\alpha_i = 0$. It might seem that $\alpha_i = 1$ are also boundaries, but this is not so because of the δ-function in the numerator. For example, in the case of three α's the projection of the hypercontour on to the real (α_1, α_2)-plane before it suffers distortion is as shown in Fig. 2.2.1. Its boundaries are $\alpha_1 = 0$, $\alpha_2 = 0$ and $\alpha_1 + \alpha_2 = 1$; the last of these is just $\alpha_3 = 0$.

We now apply the equations (2.1.22), temporarily putting primes on the α's there to distinguish them from the integration variables in (2.2.2). We get, on introducing a parameter α' corresponding to our singularity surface $S \equiv \psi = 0$,

$$\left.\begin{array}{ll} either & \alpha' = 0, \\ or & \psi = 0, \end{array}\right\} \tag{2.2.8a}$$

with a parameter $\tilde{\alpha}'_i$ corresponding to each boundary surface

$$\tilde{S}_i \equiv \alpha_i = 0,$$

$$\left.\begin{array}{llll} \textit{either} & \tilde{\alpha}'_i = 0, & \text{for each } i, \\ \textit{or} & \alpha_i = 0, & \text{for each } i. \end{array}\right\} \qquad (2.2.8b)$$

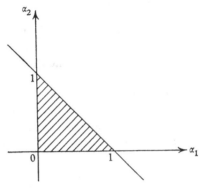

Fig. 2.2.1. The undistorted hypercontour in α_1, α_2 space shown in the real α_1, α_2 plane.

Also

$$\left.\begin{array}{l} \dfrac{\partial}{\partial k_j}[\alpha'\psi + \sum_{\nu}\tilde{\alpha}'_{\nu}\alpha_{\nu}] = 0, \quad \text{for each } j, \\[3mm] \alpha'\dfrac{\partial\psi}{\partial k_j} = 0, \quad \text{for each } j, \end{array}\right\} \qquad (2.2.8c)$$

that is

and

$$\left.\begin{array}{l} \dfrac{\partial}{\partial\alpha_i}[\alpha'\psi + \sum_{\nu}\tilde{\alpha}'_{\nu}\alpha_{\nu}] = 0, \quad \text{for each } i, \\[3mm] \alpha'\dfrac{\partial\psi}{\partial\alpha_i} + \tilde{\alpha}'_i = 0, \quad \text{for each } i. \end{array}\right\} \qquad (2.2.8d)$$

that is

We discard the possibility $\alpha' = 0$ in $(2.2.8a)$ as being trivial so that $(2.2.8a)$ and $(2.2.8c)$ together are equivalent to

$$\left.\begin{array}{l} \psi = 0, \\[3mm] \dfrac{\partial\psi}{\partial k_j} = 0, \quad \text{for each } j, \end{array}\right\} \qquad (2.2.9)$$

while $(2.2.8b)$ and $(2.2.8d)$ together are equivalent to

$$\left.\begin{array}{llll} \textit{either} & \alpha_i = 0, & \text{for each } i, \\[3mm] \textit{or} & \dfrac{\partial\psi}{\partial\alpha_i} = 0, & \text{for each } i. \end{array}\right\} \qquad (2.2.10)$$

But these equations are precisely the same as $(2.2.6)$ and $(2.2.7)$ derived from the first representation.

Third representation

In the representation (2.2.4) the α are the only integration variables. The only surface S of singularity of the integrand is $D = 0$, while the boundaries of the hypercontour are again $\alpha_i = 0$. Hence the analogues of (2.2.9) and (2.2.10) are

$$
\left.\begin{array}{ll}
& D = 0, \\
\textit{either} & \alpha_i = 0, \\
\textit{or} & \dfrac{\partial D}{\partial \alpha_i} = 0, \quad \text{for each } i.
\end{array}\right\} \tag{2.2.11}
$$

and

That these equations are essentially equivalent to (2.2.9) and (2.2.10) can be seen from (1.5.26) and (1.5.27), except that further investigation is required when $C = 0$. This matter is taken up again in §2.10.

Notice that, since D is a homogeneous function of the α_i, Euler's theorem gives

$$
\sum_i \alpha_i \frac{\partial D}{\partial \alpha_i} \propto D.
$$

Hence the first of the equations (2.2.11) is automatically satisfied when the others are. Further the homogeneity of D allows us to ignore the δ-function in the numerator of the representation (2.2.4). For if a set of α is found that satisfies (2.2.11) and such that their sum is $\nu \neq 0$, if each is divided by ν they will still satisfy (2.2.11) and now also the requirements of the δ-function.† But it is never necessary to go through this explicitly. Similar remarks apply to the Landau equations (2.2.9) and (2.2.10) derived from the representation (2.2.2).

Leading and lower-order singularities

Referring to (2.2.6) we see that in the configuration that produces the singularity, either each internal momentum is on the mass shell or the corresponding α is zero. In the latter case the momentum q_i does not appear in the other Landau equation (2.2.7b) and the corresponding mass m_i is irrelevant, so that the presence of the internal line has no effect other than on momentum conservation. Hence exactly the same singularity appears in the Feynman graph obtained from the first by contracting to points the $\alpha_i = 0$ lines. The singularity of a given graph corresponding to no $\alpha_i = 0$ lines, that is to all the lines on the mass shell, we call the *leading singularity* for the graph, while those corresponding to $\alpha_i = 0$, so that they are shared by the contracted graphs, we call the *lower-order singularities* for the graph.

---

† The case $\nu = 0$ will arise in §2.10.

If the Feynman graph were regarded as an electric circuit we should say that $\alpha_i = 0$ corresponds to the line i being short-circuited. The electric circuit analogy has been carried much further, by Bjorken (1959), T. T. Wu (1961) and Boyling (1963, 1964a). Here we note only the simplest aspects. The Landau equations (2.2.6) and (2.2.7) are formally the same as the Kirchhoff laws for an electric network identical topologically with the Feynman graph. In the electric circuit the current in the ith line is q_i and the resistance is α_i. Condition (2.2.6) tells us that unless the line is short-circuited the current must have a numerical value m_i.

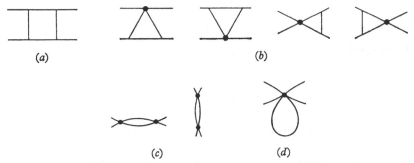

(a) (b)

(c) (d)

Fig. 2.2.2. Examples of reduced graphs (or diagrams) giving lower order singularities coming from the Feynman integral for the square diagram.

To find the complete set of lower-order singularities for a given graph, draw all the possible contracted graphs by making all possible sets of contractions of internal lines. The leading singularities for the contracted graphs are then the lower-order singularities of the original graph. Thus the lower-order singularities of the graph in Fig. 2.2.2a are the leading singularities of the graphs in Fig. 2.2.2b, each of which corresponds to one contraction, together with those in Fig.2.2.2 c corresponding to two contractions. We do not, in this case, make more than two contractions, since the Feynman integrals for the resulting graphs, such as Fig. 2.2.2d, have trivial analyticity properties.

Nature of the singularities

Landau (1959b) used the representation (2.2.2) to determine how a given Feynman integral behaves in the neighbourhood of its leading singularity. By using the representation (2.2.4), Polkinghorne & Screaton (1960b) extended the analysis to include lower-order singularities.

Suppose the Feynman graph under study has N internal lines, and consider the singularity corresponding to ν contractions where $0 \leqslant \nu < (N-1)$. We may label the α so that (2.2.11) reads

$$\left.\begin{array}{ll} D = 0, & \\ \alpha_i = 0, & i = 0, 1, \ldots, \nu, \\ \dfrac{\partial D}{\partial \alpha_i} = 0, & i = \nu+1, \nu+2, \ldots, N. \end{array}\right\} \tag{2.2.12}$$

Now perform the α_N-integration, which is trivial because of the δ-function in the numerator of the integrand in (2.2.4). This has the effect of changing $D(z, \alpha_1, \alpha_2 \ldots \alpha_N)$ to

$$D(z; \alpha_1, \alpha_2 \ldots \alpha_{N-1}, 1 - \alpha_1 - \alpha_2 - \ldots - \alpha_{N-1}) = D'(z; \alpha_1, \alpha_2 \ldots \alpha_{N-1}) \text{ say.}$$

The equations (2.2.12) become

$$\left.\begin{array}{ll} D' = 0, & \\ \alpha_i = 0, & i = 0, 1, \ldots, \nu, \\ \dfrac{\partial D'}{\partial \alpha_i} = 0, & i = \nu+1, \nu+2, \ldots, N-1. \end{array}\right\} \tag{2.2.13}$$

Now D' is not homogeneous, so that the first equation of (2.2.13) is not automatically satisfied when the others are. In fact, if the solution to the other equations is, for a given point z_r,

$$\alpha_i = \bar{\alpha}_i(z_r),$$

the equation of the surface of singularity of the Feynman integral is just
$$D'(z_r; \bar{\alpha}_i(z_r)) = 0. \tag{2.2.14}$$

In the neighbourhood of a point z_r that lies on (2.2.14) we expand D' by Taylor's theorem, retaining only the lowest terms:

$$D'(z_r; \alpha_i) = D'(z_r; \bar{\alpha}_i) + \sum_{i=1}^{\nu} (\alpha_i - \bar{\alpha}_i) \frac{\partial D'}{\partial \alpha_i}\Big|_{\alpha_i = \bar{\alpha}_i}$$
$$+ \frac{1}{2} \sum_{j,k=\nu+1}^{N-1} (\alpha_j - \bar{\alpha}_j)(\alpha_k - \bar{\alpha}_k) \frac{\partial^2 D'}{\partial \alpha_j \partial \alpha_k}\Big|_{\alpha = \bar{\alpha}} \tag{2.2.15}$$

To find the nature of the singularity, it will be sufficient to use this approximation for D' and also, although we are only concerned with some finite segment of hypercontour in the neighbourhood of $\alpha = \bar{\alpha}$, the singular part will not be changed if we let each α-integration run from $-\infty$ to ∞, provided the power $(N - 2l)$ of the denominator in

(2.2.4) is sufficiently large. Explicit integration then gives, apart from various factors, the result that the singularity is the same as in the expression

$$[D'(z_r; \bar{\alpha}_i(z_r))]^{-\gamma}, \qquad (2.2.16)$$

provided $\gamma = \frac{1}{2}[N - \nu + 1] - 2l > 0$. If $\gamma \leqslant 0$ the infinite extension of the hypercontour is not valid. However, then we may replace D' by $D' + \eta$ and differentiate the integral with respect to η a sufficient number of times to increase the power of the denominator so that the infinite extension of the hypercontour becomes valid. The result of the integration may then be integrated again with respect to η. So it emerges that when γ in (2.2.16) is a negative integer, the result (2.2.16) is to be replaced by

$$[D'(z_r; \bar{\alpha}_i(z_r))]^{|\gamma|} \log D'. \qquad (2.2.17)$$

When γ is negative but fractional (half-integer) the result (2.2.16) again holds.

For most graphs γ is negative and then the singularity is of a square root or logarithmic nature according as $N - \nu$ (the number of lines in the contracted graph or 'Landau diagram' of the singularity) is even or odd.

Care is required in applying these results to the complete amplitude because different Landau diagrams with different values of γ may have the same Landau curve, e.g. those in Fig. 2.2.3 both have $s = (4m)^2$ as leading singularity but differing γ's. In fact $s = (nm)^2$ is two sheeted only if $n = 2$.

Fig. 2.2.3. $s = (4m)^2$ graphs.

2.3 The triangle graph

We are now in a position to consider the singularities corresponding to one of the simplest Feynman graphs, that of Fig. (2.3.1a). Here the lower-order singularities are the leading singularities for the three contracted graphs in Figs. (2.3.1b, c and d). The latter singularities are known as *normal thresholds*; we shall see that they correspond exactly with those singularities whose presence we have deduced directly from unitarity (see § 1.3 and the discussion following (1.3.5)). It is not quite so simple to discover from unitarity the existence of the triangle graph's leading singularity (see § 4.11), so this singularity is known as an *anomalous threshold*. Anomalous thresholds were first investigated by Nambu (1957) and by Karplus, Sommerfield & Wichmann (1958), whose methods were the forerunners of those developed by Landau (1959 b), but which relied on the simplicity of the

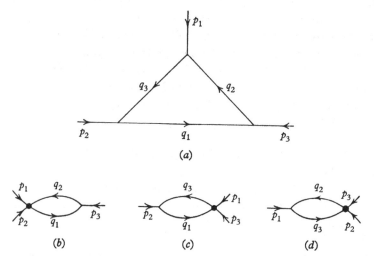

Fig. 2.3.1. The triangle graph and the three contracted (or reduced) graphs that can be obtained from it.

integral under discussion. The singularities were also studied by Källèn & Wightman (1958), who explicitly integrated a certain derivative of the Feynman integral. The problem was first approached with the techniques described here by Fowler, Landshoff & Lardner (1960).

The analysis consists naturally of two parts; the first is to find the equation of the surface Σ which *may* be a singularity of the integral, and the second to determine which parts actually are singular on a given Riemann sheet. We discuss the first part of the problem using first the (q, α)-form (2.2.6) and (2.2.7) of the Landau equations, and then again using the α-form (2.2.11). For the second part it is simplest to work with the α-representation (2.2.4) of the Feynman integral, and so we shall restrict our considerations to this.

For the graph of Fig. 2.3.1 a, with the momenta as labelled in the figure, there are three independent scalar invariants z that may be formed from the external four vectors p. We take these as

$$z_1 = p_1^2, \quad z_2 = p_2^2, \quad z_3 = p_3^2. \tag{2.3.1}$$

All other scalars that may be formed from the p may be reduced to a combination of those in (2.3.1) when use is made of the energy-momentum conservation relation

$$\sum_{i=1}^{3} p_i = 0. \tag{2.3.2}$$

Of course if each of the external particles is physical, that is on the mass shell, each of the z_i would be constant and equal to M_i^2. Here, however, we shall suppose that at least one of the p_i is off the mass shell. This will be the case when one considers a form-factor, or when more than one external line is attached to the vertex as in Fig. 2.2.2b. In the latter situation the form of the Feynman integral, and therefore of the singularity structure, is exactly the same, provided the corresponding p is regarded as the total four-momentum coming into the vertex.

First form of the Landau equations

Consider first the leading singularity, for which the Landau equations (2.2.6) and (2.2.7) read

$$q_i^2 = m_i^2 \quad \text{for each } i, \tag{2.3.3}$$

and
$$\sum_i \alpha_i q_i = 0. \tag{2.3.4}$$

The summation in (2.3.4) normally runs round a given internal loop of the graph, but here there is only one such loop and so the sum is over all i. In this summation it is necessary to make the vectors q_i point the same way all round the loop.

Multiplication of (2.3.4) by q_j yields the three simultaneous equations
$$\sum_i \alpha_i(q_i \cdot q_j) = 0 \quad (j = 1, 2, 3), \tag{2.3.5}$$

[where $(q_i \cdot q_j)$ denotes a scalar product of the four vectors]. These have a non-trivial solution for the α if and only if the 3×3 determinant

$$\det (q_i \cdot q_j) = 0. \tag{2.3.6}$$

However, energy-momentum conservation at the (q_i, q_j, p_k) vertex gives
$$p_k^2 = (q_i - q_j)^2,$$

$$z_k = m_i^2 + m_j^2 - 2(q_i \cdot q_j), \tag{2.3.7}$$

where (i, j, k) is a permutation of $(1, 2, 3)$. So $(2.3.6)$ and $(2.3.7)$ yield the equation

$$\Sigma \equiv \begin{vmatrix} 1 & -y_{12} & -y_{13} \\ -y_{21} & 1 & -y_{23} \\ -y_{31} & -y_{32} & 1 \end{vmatrix} = 0, \qquad (2.3.8)$$

where we have defined

$$y_{ij} = y_{ji} = \frac{z_k - m_i^2 - m_j^2}{2m_i m_j} = -\frac{q_i \cdot q_j}{2m_i m_j}, \qquad (2.3.9)$$

with (i, j, k) a permutation of $(1, 2, 3)$. The surface Σ is the surface in complex z_1, z_2, z_3 space on which the leading singularity can be found. We postpone the task of deciding on which parts of Σ it actually *is* found.

We may discover in a similar way the equations of the surfaces corresponding to the lower-order singularities. For example, the α_1 contraction, corresponding to the leading singularity of the graph in Fig. 2.3.1b is given by the equations

$$\left.\begin{array}{c} \alpha_1 = 0, \\ q_2^2 = m_2^2, \quad q_3^2 = m_3^2, \end{array}\right\} \qquad (2.3.10)$$

and again $(2.3.4)$, in which the first term of the sum will no longer appear because of $\alpha_1 = 0$. Multiplying $(2.3.4)$ in turn by q_2 and q_3 now yields two simultaneous equations for α_2 and α_3. The condition for these to have a solution reduces to

$$\Sigma_1 \equiv \begin{vmatrix} 1 & -y_{23} \\ -y_{32} & 1 \end{vmatrix} = 0 \qquad (2.3.11)$$

or

$$z_1 = (m_2 \pm m_3)^2. \qquad (2.3.12)$$

We may obtain the surfaces Σ_2 and Σ_3 corresponding to Figs. 2.3.1c and 2.3.1d by cyclic permutation.

Dual diagrams

One may solve the equations $(2.3.3)$ and $(2.3.4)$, and $(2.3.4)$ and $(2.3.10)$ by geometrical methods (Landau, 1959a, b; J. C. Taylor, 1960). We draw the vector diagram for internal and external momenta. Because of the energy-momentum conservation law $(2.3.2)$ the external vectors p will form a closed triangle. The squares of the lengths of the sides of this triangle are just the three z_i. The internal momenta are drawn so that energy-momentum is conserved at each vertex of the graph; hence the complete vector diagram, known in this context as the *dual diagram*, is as drawn in Fig. 2.3.2. The Landau equation (2.3.4), which says that there is a linear constraint among the three vectors q_i, requires the q_i to be drawn in a plane. Hence the whole diagram is drawn in a plane. (Unfortunately, in general, the corresponding geometrical constraint is not usually so simple as for this graph!) For the case of the leading singularity, the lengths of the

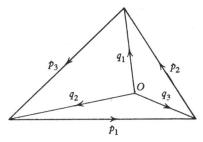

Fig. 2.3.2. The dual diagram for the triangle graph. The length of each side is given by the mass of the corresponding particle.

vectors q_i are given as the m_i. This imposes a single constraint on the shape of the external triangle; this constraint is just the equation of the desired singularity surface Σ. The constraint takes the form

$$\theta_1 + \theta_2 + \theta_3 = 2\pi, \tag{2.3.13}$$

where θ_i is the angle between the lines that represent q_j and q_k in the vector diagram. In fact

$$\cos\theta_i = -y_{jk}, \tag{2.3.14}$$

and (2.3.13) and (2.3.14) are together equivalent to (2.3.8).

The lower-order singularity Σ_1 is found by relaxing the condition that $q_1^2 = m_1^2$ and replacing it by $\alpha_1 = 0$. Then the Landau equation (2.3.4) requires q_2 and q_3 to be linearly related, that is they are parallel. Hence the point O in the vector diagram (2.3.2) lies on the side of the triangle that represents p_1, and we see at once that

$$p_1^2 = (m_2 \pm m_3)^2,$$

just as in (2.3.12).

In this discussion we have talked of the vectors as if they were Euclidean. In fact one may always pretend that the vectors in a dual diagram are Euclidean, provided it is recognised that they are liable to have complex components. In other words, the distance between two points having coordinates x_α, x'_β, where $\alpha, \beta = 1, 2, 3, 4$, is

$$[(x_1 - x'_1)^2 + (x_2 - x'_2)^2 + (x_3 - x'_3)^2 + (x_4 - x'_4)^2]^{\frac{1}{2}},$$

but the x_α, x'_β may be complex. All the usual rules of geometry apply, except that the trigonometric functions may take any values; for example, $\cos\theta_i$ in (2.3.14) need not lie between -1 and $+1$ nor even be real.

More details of dual diagram constructions are given by Okun & Rudik (1960), but see Landshoff (1960).

Second form of the Landau equations

We now derive the previous equations again, using the other form (2.2.11) of the Landau equations. According to the rules given in § 1.5 the denominator $D(p, x)$ of the Feynman integral after the internal loop momenta have been integrated out is, in the present example,

$$D = \alpha_2\alpha_3 z_1 + \alpha_3\alpha_1 z_2 + \alpha_1\alpha_2 z_3 - \left(\sum_{i=1}^{3} \alpha_i m_i^2\right)(\alpha_1 + \alpha_2 + \alpha_3). \qquad (2.3.15)$$

It is best not to simplify D by using the δ-function condition $\Sigma\alpha = 1$, since this spoils its homogeneity. The Landau equations $\partial D/\partial\alpha_i = 0$ for the leading singularity give

$$\alpha_3 z_2 + \alpha_2 z_3 - m_1^2(\Sigma\alpha) - \sum_{i=1}^{3} \alpha_i m_i^2 = 0, \qquad (2.3.16)$$

together with the two equations obtained from this by permuting the suffixes. The determinantal condition that these three equations give a non-trivial solution for the α can again be reduced to (2.3.8). The lower-order singularity surfaces Σ_1, Σ_2 and Σ_3 are obtained similarly, by putting the appropriate α's equal to zero. They correspond to the principal minors of the determinant.

Shape of the singularity surface

The equation $\Sigma = 0$ in (2.3.8) involves three complex variables y or, equivalently, the three z. It is rather difficult to have a clear picture of this six-dimensional space, so we henceforth consider the restricted problem in which one of the y (or z) is fixed at a real value. We may suppose that it is y_{23} that is fixed. From (2.3.8) we have, as the equation for Σ,

$$y_{12}^2 + y_{13}^2 + y_{23}^2 + 2y_{12}y_{13}y_{23} - 1 = 0. \qquad (2.3.17)$$

In the real (y_{12}, y_{13})-plane, that is in the plane in four-dimensional complex (y_{12}, y_{13})-space on which $\operatorname{Im} y_{12} = \operatorname{Im} y_{13} = 0$, this equation describes a conic. We distinguish three cases:

(a) $|y_{23}| < 1$. In this case the mass M_1, corresponding to the external momentum p_1, has been fixed such that each of the masses M_1, m_2, m_3 is stable against decay into the other pair:

$$M_1 < m_2 + m_3; \quad m_2 < M_1 + m_3; \quad m_3 < M_1 + m_2.$$

Then the conic is an ellipse, as drawn in Fig. 2.3.3a. The ellipse is inscribed in the square formed by the lines $y_{12} = \pm 1$, $y_{13} = \pm 1$; according to (2.3.11) these pairs of lines are just $\Sigma_3 = 0$ and $\Sigma_2 = 0$, the lower-order singularity equations.

(b) $y_{23} < -1$, so that M_1 is stable again decay into m_2 and m_3 but one or both of m_2, m_3 is unstable against decay into the other pair. Then the conic is a hyperbola, as drawn in Fig. 2.3.3b. It still has the same horizontal and vertical tangents.

(c) $y_{23} > 1$, so that M_1 is unstable against decay into the other pair, m_2 and m_3. Again we have a hyperbola, with the same horizontal and vertical tangents, as shown in Fig. 2.3.3c.

The above description only deals with the real section of the surface Σ. To draw its complex parts requires another two dimensions; to get over this difficulty it is best to use the searchline method of Tarski (1960). Imagine a line

$$y_{13} = \lambda y_{12} + \mu, \quad \lambda, \mu \text{ real}, \tag{2.3.18}$$

drawn in the real (y_{12}, y_{13}) plane and intersecting the ellipse in Fig. 2.3.3a. If λ is kept fixed and μ is increased the line moves upwards in the direction perpendicular to itself. In the process the two real intersections with the conic eventually coalesce and after that are no longer real; they have become conjugate complex points.† It is evident from (2.3.18) that the imaginary parts of their coordinates have the same or opposite signs according as $\lambda > 0$, or $\lambda < 0$. Hence we see that attached to the real arc AB of negative gradient in Fig. 2.3.3a there are two parts of complex surface running off to the top right of the figure; on one of these parts

$$\operatorname{Im} y_{12} > 0. \quad \operatorname{Im} y_{13} < 0, \tag{2.3.19a}$$

and on the other $\operatorname{Im} y_{12} < 0, \quad \operatorname{Im} y_{13} > 0.$ (2.3.19b)

Similar parts of complex surface are attached to the negative-gradient arc CD, but these parts run down towards the bottom left of the

† Note that every complex point lies on one and only one searchline, so that nothing is missed.

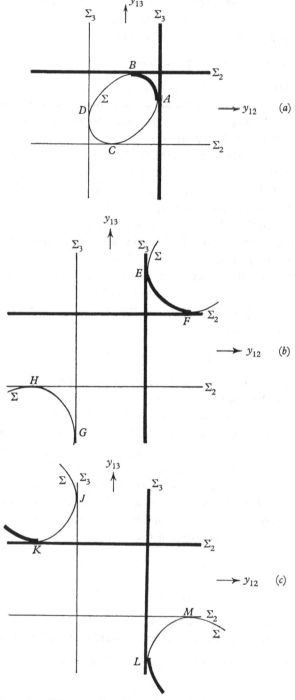

Fig. 2.3.3. The real section of the Landau surface Σ for the triangle diagram drawn in the real (y_{12}, y_{13})-plane, (a) when $|y_{23}| < 1$, (b) when $y_{23} < -1$, (c) when $y_{23} > 1$. The parts shown by heavy lines correspond to positive α (see the discussion following equation 2.3.26).

diagram. Attached to the positive-gradient arcs AC, DB, on the other hand, are parts of surface on which, either

$$\mathrm{Im}\, y_{12} > 0, \quad \mathrm{Im}\, y_{13} > 0, \qquad (2.3.20a)$$

or $\qquad\qquad\qquad \mathrm{Im}\, y_{12} < 0, \quad \mathrm{Im}\, y_{13} < 0. \qquad (2.3.20b)$

In Fig. 2.3.3b the semi-infinite positive-gradient arcs $E\infty$, $F\infty$, have attached to them complex surfaces of the type (2.3.20); the same complex surfaces are attached respectively to the arcs $G\infty$ and $H\infty$. On the other hand, the negative gradient arcs EF and GH are connected by two parts of complex surface of type (2.3.19). In Fig. 2.3.3c the situation is similar, except that the types are reversed because the gradients of the real curve are different.

Physical sheet singularity properties

Because the variables y are linearly related to the z by the definition (2.3.9), the picture in the real (z_2, z_3) plane will be very similar, except that the square formed by the horizontal and vertical tangents is now a rectangle. The location of the physical region of the variables z will depend on what lines are attached to the external vertices. However, wherever the physical region is, we have said that the appropriate integration α-space hypercontour in that region is the real undistorted one. Small imaginary parts $-i\epsilon$ are attached to the internal masses m_i to guarantee the possibility of this. Now when the value at which z_1 is fixed is real, and the α are on the original undistorted hypercontour of integration so that they are real and positive, the explicit form (2.3.15) for D gives

$$\mathrm{Im}\, D = \alpha_3 \alpha_1 \mathrm{Im}\, z_2 + \alpha_1 \alpha_2 \mathrm{Im}\, z_3 + (\Sigma \alpha_i)^2 \epsilon. \qquad (2.3.21)$$

This does not vanish for $\mathrm{Im}\, z_2$ and $\mathrm{Im}\, z_3 \geqslant 0$; hence the integral may immediately be continued throughout this region of complex (z_2, z_3)-space with no danger of the singularity $D = 0$ of the integrand forcing a distortion of the hypercontour. This proves that the physical values of the integral are boundary values† of an analytic function, the real boundary being approached from the direction $\mathrm{Im}\, z_2$, $\mathrm{Im}\, z_3 \to 0 +$. Having reached this conclusion, we may now dispense with the $-i\epsilon$ and let the internal masses be real; the $-i\epsilon$ were merely put in to tell

---

† The term 'boundary value' is a survival from the days when it was thought that the analytic continuation could not be taken across the real axis. In fact the real axis is not a natural boundary of the analytic function and, provided the discrete points of singularity are avoided, we may continue into the lower-half of the complex z_2 and z_3 planes.

us which is the correct limit on to the real boundary, to achieve the physical value of the integral.

Now the last term in (2.3.21) is missing so that we are no longer sure that $D \neq 0$ when z_2 and z_3 are real. This leads to the possibility of the integral having singularities on the real boundary, though still not of course in the region $\text{Im}\, z_2$, $\text{Im}\, z_3 > 0$. However, we can see from (2.3.15) that if the real value at which z_1 is fixed is less than $(m_2 + m_3)^2$, D will not vanish for real positive α if z_2 and z_3 are negative.† So we may continue the integral from the physical limit $\text{Im}\, z_2 = \text{Im}\, z_3 = 0+$, via the region $\text{Im}\, z_2$, $\text{Im}\, z_3 > 0$, and down to the bottom left-hand part of the real (z_2, z_3)-plane, without any singularity, $D = 0$, of the integrand forcing a distortion of the hypercontour. Next we move z_2, z_3 towards the top right of the real (z_2, z_3)-plane. Since now they are real, the situation in α-space must be symmetrical about the real hypercontour, that is, any part of the singularity surface $D = 0$ lying to one side of the hypercontour is matched by a complex conjugate part to the other side. Eventually, as we move towards the top right of the real (z_2, z_3) plane some part of $D = 0$ will come down to intersect the hypercontour, together with its complex conjugate part. Then the hypercontour cannot be distorted away from all singularities, because if we try to move it away from one part of $D = 0$ along the normal to that part, we encounter the complex conjugate part. Hence we are liable to have reached a value of z_2, z_3 for which the integral is singular; we are somewhere on the leading singularity surface $\Sigma = 0$ or one of the lower-order singularity surfaces $\Sigma_2 = 0$ and $\Sigma_3 = 0$. However, since the trapping occurs without the hypercontour having been distorted from its original position, it is only the parts of Σ, Σ_2, Σ_3 that correspond to positive α that are relevant at this point. We will prove later (following equation 2.3.26a, b) that these are just the parts drawn in heavy line in Fig. 2.3.3a. So the arc $ACDB$ and the tangents at C and D in Fig. 2.3.3a are not singularities of the integral if we approach them from the physical limit via the region $\text{Im}\, z_2$, $\text{Im}\, z_3 > 0$. If, however, we move to approach them via a different route, we might well find them to be singularities; we should then be on a different Riemann sheet.

Single variable analyticity properties

To clarify what we have discovered so far, we now fix z_2 at a real value, and fix z_1 so that $|y_{23}| < 1$. We draw again in Fig. 2.3.4a the

† This is evident for $z_1 < 0$, but for $0 < z_1 < (m_2 + m_3)^2$ some algebra is required (compare with equation (2.4.16)).

lines in Fig. 2.3.3a that are singular when we approach them by the route specified above. We have also drawn, in dotted line, a part of the ellipse that we have discovered not to be singular. Suppose first that z_2 is fixed at the real value corresponding to the line L_1 in Fig. 2.3.4a. This intersects the singular part of Σ_3 at N, given by

(2.3.12)
$$z_3 = (m_1 + m_2)^2.$$

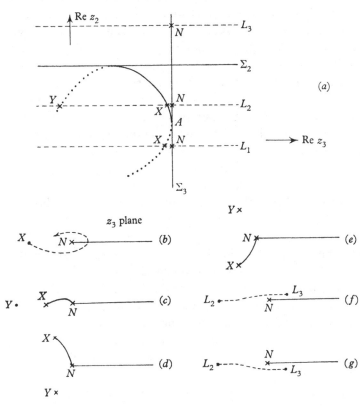

Fig. 2.3.4. (a) The real section of Σ (in the real z_2-, z_3-plane) for positive α (continuous lines); a part not singular on the physical sheet is shown by a dotted line. The broken lines L_1, L_2, L_3 denote sections with z_2 fixed and real of the complex space for which the corresponding complex z_3-plane is shown in Figs. (b), (c) and (d), (e) respectively; where (d) corresponds to L_2 moving to L_3 as in (f), and (e) to the path in (g).

This is called the two-particle *normal threshold* in the z_3-channel. It is a branch-point, so it is drawn with a cut attached to it, by convention running along the real positive axis in the complex z_3-plane, as in Fig. 2.3.4b. This convention defines the Riemann sheet that is known as the *physical sheet*. In this figure we have also drawn the point X, which is the intersection of L_1 with the non-singular arc of Σ. We assert that if we were to go from X in Fig. 2.3.4b round N, so that we pass through the cut on to another Riemann sheet, and back to X, we should now find X to be a singularity. To prove this, consider what happens when we move L_1 upwards, to the position L_2. In the z_3-plane, first X moves towards N, and coincides with it when L_1 passes through A. It is convenient to avoid this by making z_2 slightly complex, so that L_1 does not pass through A. Then X will move round N; the sense in which it moves round depends on whether we give z_2 a small positive or negative imaginary part to make L_1 avoid A. After L_1 has passed A, the intersection X moves back away from N again, and when L_1 has become L_2 we have the situation of Fig. 2.3.4c. Now, however, X represents an intersection of L_2 with the singular arc of Σ, so in Fig. 2.3.4c X represents a branch-point on the physical sheet with a cut attached. Because, in the transition from Fig. 2.3.4b to 2.3.4c, X moved round N, and so passed across the cut attached to N, we conclude that this branch-point has actually emerged from the neighbouring Riemann sheet of the branch-point N. That is, if in Fig. 2.3.4b we had gone through the cut attached to N, we should have found that X represented a singularity. This singularity has, in Fig. 2.3.4c, emerged on to the physical Riemann sheet, trailing its cut behind it. Notice that the sign given to $\text{Im } z_2$ to avoid A makes no difference to the final configuration of singularities in Fig. 2.3.4c, because there is no fixed branch-point in the z_2-plane with equation $z_2 = z_A$, where z_A is the z_2-coordinate of A (though see below). In one case the branch-point X emerges from the Riemann sheet reached from the physical sheet by going clockwise round N, and in the other from that reached by going anti-clockwise round N, but the net result is the same.

In Fig. 2.3.4c we have also drawn the point Y, which is the intersection of L_2 with the non-singular arc of Σ. If we now move L_2 upwards, X and Y approach each other and coincide when L_2 reaches Σ_2, and when L_2 has passed Σ_2 they move off in complex conjugate directions. Since neither X nor Y passes through a cut in the process, X is still singular and trailing its cut after it, while Y is still non-singular. To decide whether X or Y goes into the upper half-plane, we again have to avoid the critical configuration by temporarily giving z_2 a small imaginary part to avoid L_2 coinciding with Σ_2. For one sign of the imaginary part, X will move into the upper half plane, so that when L_2 reaches L_3 we have the situation of Fig. 2.3.4d; for the other sign X moves into the lower half plane and we have the situation of Fig. 2.3.4e. Unlike in the previous case, these two possibilities are plainly very different. This is because Σ_2 is a fixed branch-point in the z_2-plane and going round different sides of this branch-point with the variable z_2 gives different singularity structures. The two possible paths in the z_2-plane to get from L_2 to L_3 are sketched in Figs. 2.3.4f, g; here N represents the Σ_2 branch-point. Since we recall that our integral is not singular in the region $\mathrm{Im}\,z_2$, $\mathrm{Im}\,z_3 > 0$, the path of Fig. 2.3.4g must be the one that corresponds to Fig. 2.3.4d, so that Fig. 2.3.4f corresponds to Fig. 2.3.4e.

Singularity properties of the complex parts of Σ

Consider now the singularity properties of the various complex parts of Σ that sprout from the four arcs BD, AC, CD and AB in Fig. 2.3.3a. The two complex conjugate parts attached to CD stretch off to the bottom left of the figure, into the region where $\mathrm{Re}\,z_2$ and $\mathrm{Re}\,z_3 < 0$. In this region $\mathrm{Re}\,D$ does not vanish for undistorted hyper-contour of α-integration, so this region must be free of singularities on the physical sheet: we can continue our integral on to this part of Σ without being forced to distort the hypercontour. An alternative way to see that these parts of Σ are non-singular is to notice that they are connected to the real arc CD, which we already know is non-singular, by a path that lies entirely in Σ and does not cross any cut (since all the cuts have been drawn stretching off in the direction $\mathrm{Re}\,z > 0$), and we decided in § 2.1 that it normally required a cut to divide a singular part of Σ from a non-singular part. In a similar way, the two complex conjugate parts of Σ attached to the real arc AC are non-singular, since that arc is non-singular (since it lies in a region where $D < 0$ for $\alpha > 0$) and we are again able to connect every point on these parts to AC by a path lying on Σ and not passing through a cut. In any case, since AC has positive gradient one of the attached parts of complex surface lies in the region $\mathrm{Im}\,z_2$, $\mathrm{Im}\,z_3 > 0$, which we know to be free of singularity. The other lies in $\mathrm{Im}\,z_2$, $\mathrm{Im}\,z_3 < 0$, which region we can easily show is also free of singularity throughout. This is because we have already seen that for z_2, z_3 real and to the lower left-hand side of the arc AB the integration may be performed with real, undistorted hypercontour. So if we start at some real point (z_2, z_3) in this region and continue to some complex point (z_2', z_3'), the distortion of hypercontour forced on us, if any, will be just the complex conjugate of that forced on us if we continued instead by the complex conjugate path to the complex conjugate point $(\bar{z}_2', \bar{z}_3')$. Thus the analytic structure in any region Q of complex z-space is the same as that in the complex conjugate region $\bar{Q}$, provided we continue from Q to $\bar{Q}$ via the 'safe' region of the real (z_2, z_3) plane to the left of and below the arc AB.

Exactly similar considerations apply to the two complex conjugate parts of Σ attached to the arc BD; each is non-singular. However, the parts attached to the arc AB *are* singular. This is because AB is singular and they are not separated from AB by a cut. They lie in the regions

$$\mathrm{Im}\, z_3 / \mathrm{Im}\, z_2 < 0,$$

of complex z-space and are found using real searchlines of negative gradient above and to the right of AB. To get from these parts of Σ to the neighbouring, non-singular, complex parts it is necessary for either $\mathrm{Im}\, z_2$ or $\mathrm{Im}\, z_3$ to change sign; one must either pass through the cut attached to Σ_2, defined by

$$z_2\, \text{real}; \quad (m_1 + m_3)^2 \leqslant z_2 < \infty, \tag{2.3.22a}$$

or that attached to Σ_3 and defined by

$$z_3\, \text{real}; \quad (m_1 + m_2)^2 \leqslant z_3 < \infty. \tag{2.3.22b}$$

Continuation in z_1

We recall that so far we have been discussing the situation where z_1 is fixed at a real value such that figure $(2.3.3a)$ is applicable, that is $|y_{23}| < 1$.

Suppose we now vary z_1, keeping it real. Then the ellipse in Fig. $2.3.3a$ changes shape continuously. So long as we encounter no fixed branch-points in the z_1-plane, we can say that any point of Σ that does not pass through a cut retains its singularity properties. Hence, since analogy with z_2 and z_3 tells us that the only fixed branch-points in the variable z_1 is the normal threshold

$$z_1 = (m_2 + m_3)^2 \quad \text{or} \quad y_{23} = +1, \tag{2.3.23}$$

the general situation will be much the same so long as the diagram in Fig. $2.3.3a$ applies. This is the case for $|y_{23}| < 1$, or

$$(m_2 - m_3)^2 < z_1 < (m_2 + m_3)^2,$$

which is the situation considered above.

If we now decrease z_1 past the value $(m_2 - m_3)^2$ the ellipse degenerates into a pair of straight lines passing through the intersection of the normal thresholds, and then opens out into the hyperbola of Fig. 2.3.3b. In this process the complex part of Σ that was singular, the part attached to AB, has shrunk away to nothing. In fact now no complex part of Σ is singular. For, as we prove below, no part of the bottom left-hand arc of the hyperbola corresponds to positive α, and so, by arguments exactly parallel to those we used above, no part of it is singular. But each complex part of Σ is attached to some part of this arc, as can be seen by the searchline argument. For example, the part attached to the arc $F\infty$ is also attached to the arc $H\infty$.

But although no complex part of Σ is singular on the physical sheet, the real arc EF is singular if we approach it from the correct complex direction of the physical sheet. Since the complex part attached to it lies in the region

$$\operatorname{Im} z_2 / \operatorname{Im} z_3 < 0, \qquad (2.3.24)$$

and is not singular, the arc EF cannot be singular if we approach it from this direction. However, it is singular if we approach it from the direction

$$\operatorname{Im} z_2 / \operatorname{Im} z_3 > 0. \qquad (2.3.25)$$

This is because the arc corresponds to real, positive α and, as we have said before, in the region (2.3.25) of complex α-space we have an undistorted hypercontour, so that if we come down from this region on to the real plane a critical configuration of the singularity surface $D = 0$ is relevant when it occurs for real, positive α. If we go into the region (2.3.24), however, we have to distort the hypercontour away from its original position and a critical configuration occurring in its original, undistorted position need no longer be harmful.

If we now increase z_1 above the value $(m_2 + m_3)^2$ we arrive at the hyperbola drawn in Fig. 2.3.3c. The situation is now rather different, in that no real part of the real (z_2, z_3)-plane is 'safe'. Thus, although we can still see from (2.3.21) that $D \neq 0$ when the hypercontour is undistorted and $\operatorname{Im} z_2$, $\operatorname{Im} z_3 > 0$, it is no longer true that $D \neq 0$ when the hypercontour is undistorted and z_2, z_3 are real and negative. So if we cross from the region $\operatorname{Im} z_2$, $\operatorname{Im} z_3 > 0$, to $\operatorname{Im} z_2$, $\operatorname{Im} z_3 < 0$, we have to distort the hypercontour and no longer arrive at the complex conjugate situation. The reason for this asymmetry is that we have to decide which side of the normal threshold $z_1 = (m_2 + m_3)^2$ we should pass as we increase z_1, and the correct choice for the physical amplitude is in fact to go to the upper side of it. For we can see from

(2.3.21) that if we go to the lower side of it, so that $\operatorname{Im} z_1 < 0$, D is liable to vanish and force a distortion of the hypercontour. We should then not be able to arrive at the physical boundary value; Feynman's $-i\epsilon$ was put in just to avoid the need to distort the hypercontour when we calculated the physical boundary value.

Since our subsequent discussion will always be of examples where there is a 'safe' region in the real plane, we shall not discuss further the case of Fig. 2.3.3c.

Positive α

The parts of the singularity surfaces that correspond to real, positive α are drawn in heavy line in Figs. 2.3.3a and 2.3.3b. The values of the α corresponding to any point on Σ or on the lower-order surfaces Σ_i may be found explicitly from the Landau equations such as (2.3.16), but it is easier to use the dual diagrams. Thus, for the lower order surface Σ_2 we had the point O in Fig. 2.3.2 on the side p_2 of the external triangle, with

$$\alpha_2 = 0, \tag{2.3.26a}$$

$$\alpha_3 q_3 + \alpha_1 q_1 = 0. \tag{2.3.26b}$$

The point O lying on p_2 leads to the two solutions

$$z_2 = p_2^2 = (m_3 \pm m_1)^2.$$

Evidently for the $(m_3 + m_1)^2$ solution the vectors q_3 and q_1 point in opposite directions, so that (2.3.26b) leads to positive α_3 and α_1, while for the $(m_3 - m_1)^2$ solution q_3 and q_1 point in the same direction and α_3, α_1 must be of opposite signs.

Having decided in this way that the normal thresholds

$$z_2 = (m_3 + m_1)^2, \quad z_3 = (m_1 + m_2)^2$$

correspond to positive α (with respectively α_2 and α_3 zero), we can see from continuity that the arc AB in Fig. 2.3.3a and the arc EF in Fig. 2.3.3b must correspond to all the α being positive. This is because Σ and Σ_2, and similarly Σ and Σ_3, share the same values of the α at their contact, as is evident from the dual diagram. At A on the ellipse in Fig. 2.3.3a we have $\alpha_3 = 0$, so on the arc AC we have $\alpha_1, \alpha_2 > 0$, $\alpha_3 < 0$. At C we have $\alpha_2 = 0$, so on the arc CD we deduce $\alpha_1 > 0$, α_2, $\alpha_3 < 0$. Similarly, on the arc DB we find, since $\alpha_3 = 0$ at D, $\alpha_1, \alpha_3 > 0$ and $\alpha_2 < 0$. So the arc AB is the only part of the ellipse that corresponds to positive α and so, by our previous arguments, it is the only real part of the ellipse to be singular on the physical sheet. In terms of the dual diagram, Fig. 2.3.2, the condition for positive α is that the point O should be *inside* the external triangle. Hence the arc AB corresponds to

$$\theta_{12}+\theta_{23} \geqslant \pi, \quad \theta_{23}+\theta_{31} \geqslant \pi, \quad \theta_{31}+\theta_{12} \geqslant \pi, \qquad (2.3.27)$$

where the θ are defined in (2.3.14) as the angles between the lines representing the internal momenta q.

2.4 The square graph

Choice of variables

We consider now the fourth-order single-loop graph of Fig. 2.4.1, with the momenta labelled as in that figure. We suppose that each external line is on the mass shell

$$p_i^2 = M_i^2 \quad (i = 1, 2, 3, 4),$$

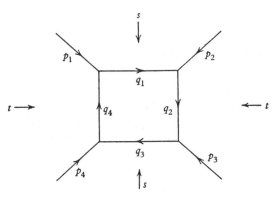

Fig. 2.4.1. The square Feynman graph.

so that the independent scalar products z of the external momenta, of which the Feynman integral is a function, may be taken as s and t,

$$\left.\begin{array}{l} s = (p_1+p_2)^2 = (p_3+p_4)^2, \\ t = (p_1+p_4)^2 = (p_2+p_3)^2. \end{array}\right\} \qquad (2.4.1)$$

The definition of these variables is the same as we adopted in (1.3.16); the third variable u defined there need not be used here, since it is dependent on s and t through the relation

$$s+t+u = \sum_i M_i^2, \qquad (2.4.2)$$

which is derived from the energy-momentum conservation relation

$$\Sigma p_i = 0, \qquad (2.4.3)$$

and the mass-shell condition.

It is useful again to define variables y, as we did in (2.3.9) for the triangle graph:

$$y_{ij} = -\frac{q_i \cdot q_j}{2m_i m_j} = y_{ji}. \qquad (2.4.4)$$

Then, since the external masses M_i are to be fixed, $y_{12}, y_{23}, y_{34}, y_{41}$, will be fixed, by virtue of the relation

$$y_{12} = \frac{M_2^2 - m_1^2 - m_2^2}{2m_1 m_2}, \qquad (2.4.5)$$

and three similar ones obtained by cyclic permutation. On the other hand, y_{24} and y_{13} vary with s, t:

$$y_{24} = \frac{s - m_2^2 - m_4^2}{2m_2 m_4}, \quad y_{13} = \frac{t - m_1^2 - m_3^2}{2m_1 m_3}. \qquad (2.4.6)$$

The surfaces of singularity

The equation of the surface Σ, in four-dimensional complex (s, t)-space, on which the leading singularity may be found, can be obtained in more than one way, as in the case of the triangle graph. The dual diagram is drawn in Fig. 2.4.2. The external vectors p form a closed quadrilateral, because of the energy-momentum conservation condition (2.4.3), and each is of a definite length M which we have fixed. To get the leading singularity each internal line q must be put on the mass shell

$$q_i^2 = m_i^2, \qquad (2.4.7)$$

and also there must be a set of parameters α such that

$$\Sigma \alpha_i q_i = 0, \qquad (2.4.8)$$

as in (2.2.6) and (2.2.7). The latter condition, being a linear relation among four vectors, requires those four vectors to lie in a three-dimensional subspace of the four-dimensional Lorentz space. This means that the whole dual diagram must be drawn in a three-dimensional space. This places a single constraint on the shape of the quadrilateral $ABCD$, which yields an equation relating the lengths AC and BD of its diagonals. These lengths are just $\sqrt{s}$ and $\sqrt{t}$ respectively, and the constraint is just the desired equation of the surface Σ. It may thus be found geometrically.

Alternatively, we may multiply (2.4.8) by each of the q_j in turn:

$$\sum_i \alpha_i q_i \cdot q_j = 0 \quad (j = 1, 2, 3, 4), \tag{2.4.9}$$

and write down the condition that these four simultaneous linear equations have a solution with the α not all zero. If we use (2.4.7) and the definitions (2.4.5 and 2.4.6), this condition reads

$$\begin{vmatrix} 1 & -y_{12} & -y_{13} & -y_{14} \\ -y_{12} & 1 & -y_{23} & -y_{24} \\ -y_{13} & -y_{23} & 1 & -y_{34} \\ -y_{14} & -y_{24} & -y_{34} & 1 \end{vmatrix} = 0 \tag{2.4.10}$$

which is the desired equation for Σ.

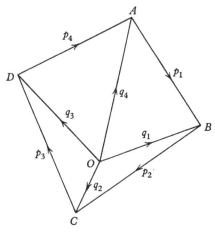

Fig. 2.4.2. The dual diagram for the square Feynman graph that is drawn in Fig. 2.4.1.

Finally, we may use the explicit form for the denominator function D that is to be inserted in the integral (2.2.4). According to the rules of § 1.5, this is

$$D = \alpha_2 \alpha_4 s + \alpha_1 \alpha_3 t + \alpha_4 \alpha_1 M_1^2 + \alpha_1 \alpha_2 M_2^2$$
$$+ \alpha_2 \alpha_3 M_3^2 + \alpha_3 \alpha_4 M_4^2 - (\Sigma \alpha_i m_i^2)(\Sigma \alpha). \tag{2.4.11}$$

The equations, $\partial D / \partial \alpha = 0$, of (2.2.11) yield four simultaneous linear equations for the α, and the condition that these have a non-trivial solution may again be reduced to the form (2.4.10).

The lower-order singularities, corresponding to one of the internal lines being contracted, are the leading singularities for four triangle graphs in Fig. 2.4.3 a–d. Those corresponding to two α being contracted are the 'normal threshold' graphs in Fig. 2.4.3 e, f. The other lower-order singularities corresponding to two contractions would only be of interest if the external masses M_i were variables, for example the (α_1, α_2) contraction would give a surface with equation

$$M_2^2 = (m_3 \pm m_4)^2.$$

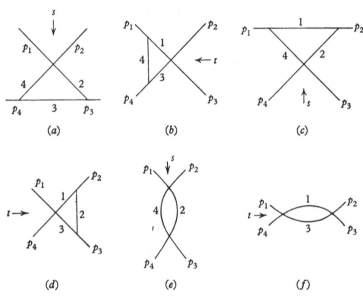

Fig. 2.4.3. Reduced diagrams obtained by contracting lines in the square Feynman graph. These give lower order singularities of the Feynman integral for the square graph.

We already know, from §2.3, what are the equations for the surfaces $\Sigma_1, \Sigma_2, \Sigma_3, \Sigma_4$, corresponding to the triangle singularities. They are just obtained from the four leading minors of the determinant in (2.4.10) (compare (2.3.8)), namely

$$\begin{vmatrix} 1 & -y_{23} & -y_{24} \\ -y_{23} & 1 & -y_{34} \\ -y_{24} & -y_{34} & 1 \end{vmatrix} = 0, \qquad (2.4.12)$$

and three similar equations. We also know, as in (2.3.12), that the surfaces Σ_{24}, Σ_{13} corresponding to the contracted graphs of Fig. 2.4.3 e, f have equations

$$\left. \begin{aligned} y_{24} &= \pm 1, \quad \text{or} \quad s = (m_2 \pm m_4)^2, \\ y_{13} &= \pm 1, \quad \text{or} \quad t = (m_1 \pm m_3)^2. \end{aligned} \right\} \qquad (2.4.13)$$

The singularity structure

Once again the physical values of the Feynman integral are to be found by attaching an imaginary part $-i\epsilon$ to each internal mass m_i^2, and then integrating over the real, undistorted α-hypercontour. We have, from (2.4.11),

$$\operatorname{Im} D = \alpha_2 \alpha_4 \operatorname{Im} s + \alpha_1 \alpha_3 \operatorname{Im} t + (\Sigma \alpha_i)^2 \epsilon \qquad (2.4.14)$$

on the undistorted hypercontour. Hence, just as in the discussion of (2.3.21), we may immediately continue the integral from the real, physical boundary throughout the region $\operatorname{Im} s$, $\operatorname{Im} t > 0$, without danger of D vanishing on the undistorted hypercontour and so forcing a distortion. Provided that we have chosen the fixed values of the external masses suitably, we also find that D does not vanish on the undistorted hypercontour for s, t in some region R of the *real* (s, t) plane. Then we may also continue the integral from the region $\operatorname{Im} s$, $\operatorname{Im} t > 0$ on to R. Then, exactly as in the discussion of §2.3, we know:

(*a*) That the boundary of R is composed of those parts of Σ, Σ_i, Σ_{13}, Σ_{24} that correspond to real, positive α.

(*b*) That complex conjugate regions of complex (s, t) space have similar analytic properties. We recall that this is because, if we continue from R to some point (s, t) by some path P and are forced to make some distortion of the hypercontour, we must make the complex conjugate distortion if we continue to the complex conjugate point $(\bar{s}, \bar{t})$ along the conjugate path $\bar{P}$, and so end up with just the complex conjugate situation.

The condition for the region R to exist is that we fix the external masses M_i such that each M_i is stable against decay into the pair of internal particles at the corresponding vertex of the graph. So we require

$$\left.\begin{aligned} M_2 &< m_1 + m_2,\ \text{etc.,} \\ y_{12}, y_{23}, y_{34}, y_{14} &< 1. \end{aligned}\right\} \qquad (2.4.15)$$

or

That R does indeed exist when (2.4.15) holds may easily be seen by re-writing the explicit form (2.4.11) for D as (Karplus, Sommerfield & Wichmann, 1959)

$$\begin{aligned} D = {} & \alpha_2 \alpha_4 [s - (m_2 - m_4)^2] + \alpha_1 \alpha_3 [t - (m_1 - m_3)^2] \\ & - (\alpha_1 m_1 + \alpha_3 m_3 - \alpha_2 m_2 - \alpha_4 m_4)^2 - \alpha_4 \alpha_1 [M_1^2 - (m_1 + m_4)^2] \\ & - \alpha_1 \alpha_2 [M_2^2 - (m_1 + m_2)^2] - \alpha_2 \alpha_3 [M_3^2 - (m_2 + m_3)^2] \\ & - \alpha_3 \alpha_4 [M_4^2 - (m_3 + m_4)^2], \end{aligned} \qquad (2.4.16)$$

from which we see that R contains at least the region

$$s < (m_2 - m_4)^2, \quad t < (m_1 - m_3)^2. \tag{2.4.17}$$

The details of which parts of Σ, Σ_i, correspond to real, positive α, and so can provide the boundary of R, depend on the values at which the external masses M_i are fixed. If these do not provide any parts of the boundary, we have seen in §2.3 that the normal thresholds

$$s = (m_2 + m_4)^2,$$

which is a part of Σ_{13} (see 2.4.13) and

$$t = (m_1 + m_3)^2,$$

which is a part of Σ_{24}, do correspond to real positive α and so will provide the whole boundary.

We suppose here that the masses M_i are fixed such that each of the y in (2.4.15) is less than one in modulus and, further, such that the sum of any two of the corresponding angles is less than π. We recall that we define θ, as in (2.3.14), by

$$\cos \theta = -y.$$

The second of these conditions guarantees that none of the lower-order triangle surfaces Σ_i of Fig. 2.4.3 a–d has a part corresponding to real, positive α. This we discovered in §2.3; only when the line L in Fig. 2.3.4 intersected the arc AB in Fig. 2.3.3 a did we get a triangle singularity on the physical sheet, and the condition for this was (2.3.27).

With these constraints, we find (Tarski, 1960) that the section in the real (s, t) plane of the quartic surface, Σ, whose equation is (2.4.10), has the appearance of Fig. 2.4.4. It consists of four infinite arcs $\gamma_1, \gamma_2, \gamma_3, \gamma_4$ and a closed loop γ_5. Its horizontal and vertical tangents are the Σ_i, the surfaces corresponding to one contraction, and its asymptotes are Σ_{13}, Σ_{24}, which correspond to two contractions. The precise ordering of the lines Σ_i in the figure depends on the relative magnitudes chosen for the M_i.

When, as in the case we are considering, none of the parts of Σ_i corresponds to real, positive α, no part of any γ to either side of its contact with a Σ_i can correspond to real positive α. This follows from the continuity of the values of the α round an arc, and the fact that at a contact, Σ and Σ_i share the same values of the α (with one of them zero). This latter fact may readily be seen from the dual diagram of Fig. 2.4.2. At the contact of Σ and Σ_1, for example, the dual diagram must be drawn with all the lines on the mass shell and also $\alpha_1 = 0$; then actually O, A, C, D are coplanar. At the infinite intersection of the asymptotes with the curve the values of the α on Σ and Σ_{13} (or Σ_{24}) are again shared (the dual diagram is drawn with A, O, C (or B, O, D) collinear, so that B, D (or A, C) go off to infinity). Hence, again by the continuity argument, the arc γ_1, being asymptotic to both the normal

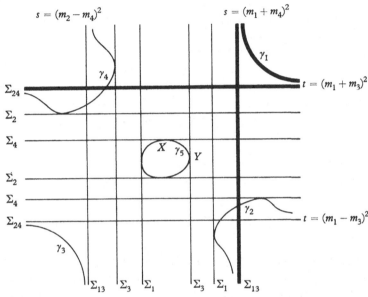

Fig. 2.4.4. The real section of the surfaces of singularity for the square Feynman graph. The curves correspond to the leading singularities, and the straight lines to lower order singularities from reduced diagrams. Heavy lines denote curves that are singular for positive α when conditions stated in the text are satisfied.

thresholds each of which corresponds to positive α (with two of them zero), corresponds to all the α real and positive. But this arc is the other side of the normal thresholds from the region R, so it is the normal thresholds that provide the boundary of R.

Just as in the discussion of (2.3.18), we may use the searchline method to discover the complex parts of Σ. Each of the arcs γ_1, γ_2, γ_3, γ_4, γ_5 has sprouting from it parts of complex surface. However, since the latter four arcs lie in R and so are not singular on the physical sheet, the parts of complex surface attached to them cannot be singular, not being divided from them by cuts. Also, the two complex conjugate parts of complex surface attached to γ_1 are not singular, because those parts are also joined to the non-singular arc XY of γ_5. These parts lie in the region

$$\text{Im}\,s/\text{Im}\,t < 0,$$

so if we approach γ_1 from these directions we must also find it to be non-singular. However, because γ_1 corresponds to positive α's, if we approach from the opposite directions,

$$\text{Im}\,s/\text{Im}\,t > 0,$$

we expect it to be singular, just as in the discussion after (2.3.25) of the arc EF in Fig. 2.3.3b.

Hence, for the particular choice of masses M_i that we have been considering, there are no complex singularities on the physical Riemann sheet. This property leads to a Mandelstam representation (1.3.33) for the graph, though only the first of the double integrals there is needed, since this graph has no normal threshold in the u-channel, so that $\quad \rho_{tu} = \rho_{us} = 0.$

The analytic properties when the M do not satisfy the conditions we have imposed in this section are discussed in § 2.7.

2.5 Single variable dispersion relations and physical-region singularities

For both the triangle graph and the square graph we discovered a region R in the real z-plane for which the Feynman integral was defined with undistorted hypercontour of α-integration. The existence of such a region R is useful for two reasons. First, it simplifies the discussion of the analytic properties of the Feynman integral as a function of several complex variables z, in that analytic continuations out of R along complex-conjugate paths P, $\bar{P}$ in z-space just lead to complex conjugate distortions of the integration hypercontour so that if we can prove analyticity in one region we are guaranteed it in the complex-conjugate region. Secondly, the knowledge that the Feynman integral is analytic in the real region R enables one, as we show below, very simply to deduce that the integral is actually analytic in a larger region, which is sufficient to allow a single-variable dispersion relation. So if the intersection R_0 of the regions R for all the Feynman graphs that contribute to a given process is non-empty,† a single-variable dispersion relation may be derived that is valid for each of those graphs. If the perturbation series converged uniformly, it would then

† R_0 is sometimes called the Symanzik region, or the Wu region.

be certain that the complete transition amplitude for the given process satisfied the same dispersion relation. However, since there certainly is not uniform convergence, the argument can only be regarded as one of plausibility.

The intersection R_0 of the regions R for a complete set of graphs was first investigated by Nambu (1957), Symanzik (1958), J. C. Taylor (1960) and Eden (1960 a). More powerful methods have recently been used by Logunov, Todorov & Chernikov (1962), T. T. Wu (1961) and Boyling (1963, 1964 a), who have exploited the analogy between a Feynman diagram and an electric circuit. To make our discussion self-contained, we here confine our discussion to the less sophisticated techniques. The general method is called 'majorisation' and is directed towards showing that all diagrams for a given process lead to Feynman integrals that are regular analytic functions in a domain R_0 whose extent is determined from one or more simple diagrams.

Vertex function in equal-mass theory

For definiteness, we consider the form factor of a particle of mass m, in a theory that contains no other particles. So we must consider all Feynman graphs that contribute to the process of Fig. 2.5.1, where

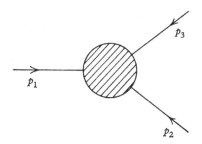

Fig. 2.5.1. The vertex diagram that corresponds to all Feynman graphs with the given external lines.

$p_2^2 = p_3^2 = m^2$, each internal line in any graph corresponds to mass m, and each integral is studied as a function of the single variable $z = p_1^2$. Of course, having only one variable simplifies the problem greatly; for each graph, the region R is just a segment of the real axis in the complex z-plane.

According to § 1.5, the denominator function D of the representation (2.2.4) of the Feynman integral is given by

$$D = -(\mathbf{Bp})^T . \mathbf{X}(\mathbf{Bp}) + (\mathbf{p}^T . \mathbf{\Gamma p} - \sigma) C, \qquad (2.5.1)$$

where the matrices $\mathbf{B}, \mathbf{X}, \boldsymbol{\Gamma}$ are obtained by writing the function ψ, which is a quadratic in the internal momenta q_i of the graph,

$$\psi = \sum_i \alpha_i(q_i^2 - m_i^2), \qquad (2.5.2)$$

in the form $\psi = \mathbf{k}^T . \mathbf{A} \mathbf{k} - 2\mathbf{k}^T . \mathbf{B} \mathbf{p} + (\mathbf{p}^T . \boldsymbol{\Gamma} \mathbf{p} - \sigma), \qquad (2.5.3)$

with $\mathbf{X} = \operatorname{adj} \mathbf{A}, \quad C = \det \mathbf{A}, \quad \sigma = \sum_i \alpha_i m_i^2. \qquad (2.5.4)$

By definition, R_0 is that part of the real axis in the complex z-plane for which no D for any graph vanishes on the undistorted α-hyper-contour, that is with all the α real and positive.

According to the rules given in §1.5 for writing down D for any graph, C is a sum of products of the α. Hence C is positive on the un-distorted hypercontour. Another way to see this is to equate the expressions for ψ in (2.5.2) and (2.5.3) for the particular case $\mathbf{p} = 0$, $m_i = 0$. Then

$$\sum_i \alpha_i q_i^2 = \mathbf{k}^T . \mathbf{A} \mathbf{k}, \quad [\mathbf{p} = 0, \, m_i = 0]. \qquad (2.5.5)$$

If the q_i were Euclidean (rather than Lorentz) vectors, the left-hand side of (2.5.5) would, for positive α, be a positive-definite quadratic form in the k, and therefore also the right-hand side. But the condition for the right-hand side to be positive definite for Euclidean momenta is just that $C = \det \mathbf{A}$ be positive and that all its principal subdetermi-nants are positive. Since C is a function only of the α, this result is independent of our choosing to prove it by making the momenta Euclidean and taking $\mathbf{p} = 0$, $m_i = 0$.

In the same way, we equate (2.5.2) and (2.5.3) when $\mathbf{k} = 0$, and get then

$$\sum_i \alpha_i(q_i^2 - m_i^2) = \mathbf{p}^T . \boldsymbol{\Gamma} \mathbf{p} - \sigma, \quad [k = 0]. \qquad (2.5.6)$$

But each q_i is a linear combination of the various k and p_2, p_3. Suppose that we have chosen to run p_2 and p_3 through the graph (and out along the external line p_1) such that each internal momentum q_i depends on at most one of p_2, p_3. Then, since we are discussing the case

$$p_2^2 = p_3^2 = m^2,$$

when $\mathbf{k} = 0$ each q_i^2 is either 0 or m^2. Hence, as we are also taking each $m_i^2 = m^2$, the left-hand side of (2.5.6) is negative for positive α, and so also the right-hand side. Thus the second term in the expression (2.5.1) for D is negative.

If the vectors p are Euclidean the first term in D is also negative.

We see this in the following way: the quadratic forms

$$\mathbf{x}^T . \mathbf{Ax}, \quad \mathbf{x}^T . \mathbf{Xx} \qquad (2.5.7)$$

may be simultaneously diagonalised and we already know from (2.5.5) that the first of these is positive definite, so that $\det \mathbf{A}$ and all the leading minors of the diagonalised matrix $\mathbf{A}$ must be positive. Hence since $\mathbf{X} = \mathrm{adj}\,\mathbf{A}$, we know also that $\det \mathbf{X}$ and all the leading minors of the diagonalised matrix $\mathbf{X}$ are also positive. So the positive definiteness of the first quadratic form in (2.5.7) implies a similar property for the second. So if the p are Euclidean $(\mathbf{Bp})^T . \mathbf{X}(\mathbf{Bp})$ is positive and the first term in the expression (2.5.1) for D is negative. So, for Euclidean p and positive α, the whole expression for D is negative.

Now the condition that the p can be Euclidean is

$$\sqrt{p_1^2} \leqslant \sqrt{p_2^2} + \sqrt{p_3^2}; \quad \sqrt{p_2^2} \leqslant \sqrt{p_1^2} + \sqrt{p_3^2}; \quad \sqrt{p_3^2} \leqslant \sqrt{p_1^2} + \sqrt{p_2^2}; \quad (2.5.8)$$

so when $p_2^2 = p_3^2 = m^2$, they are Euclidean when $z = p_1^2$ satisfies

$$0 \leqslant z \leqslant 4m^2,$$

and so D is certainly negative for undistorted hypercontour and z in this part of the real axis. But D is linear in z, and according to the rules of § 1.5 the coefficient of z is a sum of products of α and so positive. Hence the region in which D does not vanish for undistorted hypercontour is

$$-\infty < z < 4m^2. \qquad (2.5.9)$$

At $z = 4m^2$ there is the normal threshold, so (2.5.9) is the region R_0 we set out to find. It should be stressed that in this analysis we have made crucial use of the equal-mass condition; the region R_0 does not always extend up to the normal threshold.

Other amplitudes

In the case of a scattering amplitude we must consider two variables s, t as in the example of § 2.4. In that example the region R in the real (s, t)-plane extended to $-\infty$ in both variables, but for a more general graph we know that this cannot be so. This is because a general graph possesses normal thresholds in the u-channel, and u is linearly related to s, t by (2.4.2). So if the intersection R_0 of the regions R for the set of scattering graphs exists, we expect it to lie somewhere within the triangle in the real (s, t)-plane formed by the normal thresholds. This is

$$s \leqslant 4m^2, \quad t \leqslant 4m^2, \quad 4m^2 - s - t = u \leqslant 4m^2, \qquad (2.5.10)$$

in the equal-mass case. By arguments exactly parallel to those we have used for the vertex graphs, we may show that R_0 at least contains the part of the interior of this triangle that corresponds to Euclidean external vectors p, namely

$$s > 0, \quad t > 0, \quad 4m^2 - s - t = u > 0.$$

Some further work (Eden, 1960a; T. T. Wu, 1961) reveals in the equal mass case that R_0 is actually the whole region (2.5.10).

When the equal-mass restriction is relaxed, T. T. Wu (1961) has shown that a region R_0 still exists for scattering, provided no external mass exceeds $\sqrt{2}$ times the smallest internal mass. His analysis also covers multiparticle amplitudes, provided not more than ten external particles are involved. Boyling (1963, 1964a) has shown that the restriction on the external masses may be relaxed somewhat, provided it is replaced by a conservation law. He has been able to apply the analysis to pion-nucleon interactions, taking the masses at their physical values and using baryon conservation and the pseudo-scalar nature of the pion. The results of Wu and Boyling using more powerful methods of majorisation are that the region R_0 for a given process is bounded by leading singularities corresponding to some of the simplest graphs.

Single-variable dispersion relations

The real region R_0 in which the Feynman integrals for a given process are well defined with undistorted hypercontour, and therefore analytic, may readily be extended in two ways (T. T. Wu 1961):

(a) If $D \neq 0$ for α real, positive and $z \in R_0$, D similarly will not vanish if the z are made complex, such that $\text{Re}(z) \in R_0$. So all the Feynman integrals are analytic in the 'tube' in the space of the complex variables z whose real section is R_0.

(b) Suppose the real point $z_r = Z_r$ is in R_0 and define a line L through Z_r in the real subspace of the complex variables z_r:

$$L: z_r = Z_r + \lambda_r x. \tag{2.5.11}$$

Here the numbers λ_r are fixed; they determine the direction of L. Different points on L are given by different values of the real parameter x. Because D is linear in the z_r, at the point on L determined by a given value of x it takes a value that may be written

$$D(x) = F(\alpha) x + G, \tag{2.5.12}$$

where $F(\alpha)$ is a function of the α only. Because the point $x = 0$ (that is the point $z = Z$), is in R_0, the function G does not vanish when the α

are on the undistorted hypercontour. Now replace x by the complex variable ζ

$$\zeta = x + iy.$$

Then, when the α are on the undistorted hypercontour, so that they are real,

$$\operatorname{Im} D(\zeta) = yF, \quad \operatorname{Re} D(\zeta) = xF + G. \tag{2.5.13}$$

We see that $D(\zeta)$ cannot vanish when $\operatorname{Im} \zeta \neq 0$ and the α are on the undistorted hypercontour because, when $y \neq 0$, $\operatorname{Im} D = 0$ implies $F = 0$, and if then also $\operatorname{Re} D = 0$ it would be necessary that $G = 0$, which we have discovered not to be the case. So all the Feynman integrals for the given process are well defined with undistorted hypercontour throughout the complex part of the ζ-plane, and also for those real values x of ζ such that the point z defined by (2.5.11) lies in R_0. Hence the integrals are analytic in the ζ-plane, with cuts along the real axis, except for a cut-free gap corresponding to the intersection of L with R_0.

These analytic properties are just what we need to allow a dispersion relation in the variable ζ (see §1.4). So for the vertex part V in equal-mass theory, where we worked with only one variable z and found in (2.5.9) that R_0 extends from $z = -\infty$ to $z = 4m^2$, we have

$$V(z) = \frac{1}{\pi} \int_{4m^2}^{\infty} \frac{\operatorname{disc} V(\zeta)\, d\zeta}{z - \zeta}. \tag{2.5.14}$$

Here we have supposed that $V(z) \to 0$ at infinity, so that no subtraction is necessary.

In exactly the same way, for elastic scattering in the equal mass case the existence of the region R_0 of (2.5.10) leads directly to single-variable dispersion relations (Eden, 1960b). Now there are two independent variables and so we have an infinity of different choices of the line L in (2.5.11). For example, if we take L to be

$$t = t_0$$

where t_0 is a constant, satisfying

$$-4m^2 < t_0 < 4m^2$$

so as to make L pass through R_0, we obtain a fixed-t dispersion relation. Similarly, there are dispersion relations in which s, u or a suitable linear combination of the variables is fixed.

A partial-wave dispersion relation can also be derived. The partial-wave amplitude $a_l(s)$ is defined in (1.3.36) by

$$a_l(s) = \frac{1}{2} \int_{-1}^{1} d(\cos\theta)\, \tilde{F}(s, \cos\theta)\, P_l(\cos\theta) \tag{2.5.17}$$

where
$$\cos\theta = 1 + \frac{2t}{s - 4m^2}. \tag{2.5.18}$$

We assert that when the integral (2.5.17) is continued from physical values of s, where the contour in the complex $\cos\theta$ plane is by definition real, into the complex s-plane no distortion of the contour is required for any complex value of s. To see this, note that for l an integer P_l is a polynomial and so has no singularity to force a distortion. That no singularity of $\bar{F}$ forces a distortion arises from the fact that when $\cos\theta$ is fixed at a value Z_0, where $-1 \leqslant Z_0 \leqslant 1$, (2.5.18) represents the equation of a line L that passes through the region R_0 of (2.5.10) and so contains no complex points of singularity. So for complex s no singularity of the integrand in (2.5.17) crosses the real undistorted contour and $a_l(s)$ can only have singularities at real values of s. If the behaviour of a_l as $|s| \to \infty$ is suitable, this guarantees the validity of a dispersion relation to all orders of perturbation theory in the equal mass case. For unequal masses the relation between angle and momentum transfer does not take the simple form (2.5.18) and this method cannot be used to establish the validity of a partial-wave dispersion relation.

Similar methods of deriving dispersion relations may be applied to multiparticle amplitudes. These are considerably more complicated, if only because there are so many more independent variables (see § 4.3). For example, for the two-particle → three-particle amplitude it is natural to work with ten scalar products of the four-momenta, though only five of these are independent (just as of the variables s, t, u for scattering only two are independent). It may be shown that there are no simple single-variable dispersion relations for which four of the five independent variables are fixed at physical values (Landshoff & Treiman, 1962). This is because one of two things happens, according to which of the variables are fixed: either

(i) the analogues of the right-hand and left-hand cuts in Fig. 1.3.4 overlap, with the physical limit to be taken *between* them. This means that any continuation away from the physical limit involves pushing the cuts out of the way and so exposing the Riemann sheet below, on which one expects to find all sorts of complex singularities;

or

(ii) the normal threshold cuts do not overlap, but complex singularities occur on the physical sheet. These are produced by triangle graphs and probably also by other graphs.

Nevertheless, because there does exist a region R_0 for the two-particle $\to$ three-particle amplitude, there are some simple dispersion relations for this amplitude (Branson, Landshoff & Taylor, 1963). These are obtained again by taking a line L through R_0, so that only real singularities arise for the corresponding choice of variable. The trouble is that this corresponds to all the primitive variables varying together, so that the dispersion relations do not seem easy to use.

For amplitudes involving more than five particles yet further complications arise. The constraints among the primitive variables, of the type (1.2.12), that arise from energy-momentum conservation are linear. But when there are many particles one also has quadratic constraints, arising from the dimensionality of space-time. This is explained in § 4.3. Its effect is that, if one chooses a set of independent scalar-product variables and expresses the remaining variables in terms of them, square-root branch points are introduced, which must be reflected in any dispersion relation.

Physical region singularities

We now show that the only singularities in the physical regions are those such that, in the diagrams for which the singularities are the leading singularities, all the internal momenta q_i (as well as the external momenta) take physical values in the critical configuration (which is the configuration for which all variables satisfy the Landau equations). We then use this result to show that the only singularities of two-particle–two-particle scattering amplitudes in the physical regions are the normal thresholds. In physical regions the physical unitarity condition holds and can be used to derive these results. The proof by this method is due to Eden (1960a) and Landshoff (1962). A stronger result was later proved by Coleman & Norton (1965): not only do the internal momenta q_i have to take physical values to give a physical-region singularity; they also must be such that, if each vertex of the Feynman graph be regarded as representing a point interaction, these points can occur at physical values of the space-time coordinates.

First, it is easy to prove that the q_i must be real for a physical-region singularity, and that further the α must be real and positive. For if we use the (k, α) representation (2.2.2) of the Feynman integral, the physical values of the integral are given by integrating over the undistorted (k, α)-hypercontour. The possibility of this is guaranteed by Feynman's addition of $-i\epsilon$ to each internal mass. The presence of the

$-i\epsilon$ keeps the physical regions clear of singularities. As ϵ is allowed to tend to zero the normal thresholds and other possible physical-region singularities come on to the real axis, but the hypercontour can remain undistorted till the last moment. So the physical-region singularities must correspond to real, positive α and real k. Since the external momenta p are real in the physical regions, and each q is a linear combination of the k and the p, this means that the q are real.

Now if a singularity is the leading singularity for a graph, we must also have the Landau equations (2.2.6a) and (2.2.7b):

$$q_i^2 = m_i^2, \tag{2.5.19}$$

$$\sum_{\substack{\text{any} \\ \text{loop}}} \alpha_i q_i = 0. \tag{2.5.20}$$

The reality of the q, together with the on-mass-shell condition (2.5.19) implies that they can be regarded as the momenta of physical particles. Now we see that it is consistent to identify $\alpha_i m_i$ with a constant multiple of the proper time of flight of the particle i. The positive-α condition guarantees that each particle moves forward in time, while (2.5.20) guarantees that if we follow two different routes, by different chains of internal lines, between two vertices in a Feynman graph, the total space-time displacement associated with either route is the same. Hence we see that the leading singularity of any graph only occurs in the physical regions if the internal vertices can be regarded as point interactions occurring successively at real space-time points and with the associated particles having physical momenta.

Now if all the particles in the theory are stable,† at least four of them must be involved in an interaction if all their momenta are to be physical. So for two-particle → two-particle scattering the only leading singularities that can occur in the physical region are those for graphs of the structure of Fig. 2.5.2a. These are characterised by the initial-state particles entering the same vertex, and the final-state particles leaving the same vertex. This class of graphs includes the normal-threshold graphs, and in fact only the latter can be singular in the physical regions. To see this, notice that the graphs of the class of Fig. 2.5.2a correspond to Feynman integrals that are functions only of the total momentum p in the initial (or final) state, rather than of the momenta of the two individual particles in the state. By con-

---

† Unstable particles, as we see in § 4.9, contribute to Landau singularities in the same way as stable ones, except that their masses are given complex values. This displaces the corresponding singularities off the real axis, so they do not occur in the physical region.

sidering dual diagrams, it is easy to see that in the case of normal thresholds the internal momenta in the critical configuration are parallel to p, while for all other graphs of the class they are not. This implies that only the normal thresholds can correspond to positive α and so be physical-region singularities. This we see as follows: in the notation of (2.5.3) the Landau equations (2.5.20) read

$$\mathbf{Ak} = \mathbf{Bp}. \qquad (2.5.21)$$

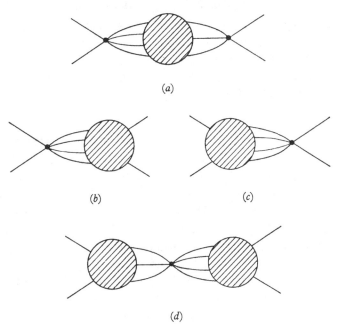

Fig. 2.5.2. Diagrams, for equal-mass theory, showing the general structure of (a) any Feynman graph that is singular on the physical sheet; (b), (c) vertex graphs; (d) a product of vertex graphs, having only normal thresholds on the physical sheet.

This can only produce a solution for the momenta k not in the space spanned by the momenta p (in the present case there is just one p) if

$$C = \det \mathbf{A} = 0. \qquad (2.5.22)$$

But we know that C is a sum of products of the α, and so cannot vanish for positive α.

A corollary of this result is that, in the equal-mass theory, the only singularities of vertex graphs on the physical Riemann *sheet* are the normal thresholds. For the vertex graphs have the same internal structure as the scattering graphs of Fig. 2.5.2 b, c. Hence they can have no singularities other than normal thresholds in the physical regions for scattering. In the s-channel the physical region is $s \geqslant 4m^2$. On the other hand, according to (2.5.9) the region R_0 extends from $-\infty$ up to $4m^2$, and in this region the vertex graphs can have no singularities, because for a vertex graph D has the structure

$$D = f(\alpha)\, s + K(\alpha, m^2), \qquad (2.5.23)$$

which does not vanish on the undistorted hypercontour if we continue from R_0 into Im $s \neq 0$. Hence the only singularities of the vertex graphs in the equal-mass case are the normal thresholds.

Similar results apply to the class of graphs of Fig. 2.5.2d, because each such graph represents an integral that factorises into the product of two integrals, one corresponding to a graph of the class of Fig. 2.5.2b, and the other of Fig. 2.5.2c.

2.6   Scattering amplitude as a function of two variables

We now briefly review techniques for investigating the whole class of Feynman graphs that contribute to a two-particle–two-particle scattering amplitude. The aim of this investigation (Eden, 1960b, c, d, 1961; Landshoff, Polkinghorne & Taylor, 1960, 1961) was to show that each graph had, on the physical Riemann sheet, only real singularities. This property is necessary for the Mandelstam representation (1.3.33) to be valid. The proof of the property is, however, incomplete (Eden, Landshoff, Polkinghorne & Taylor, 1961b); the reasons for this are described in § 2.8.

The analysis proceeds by induction; when any given graph is being considered, it is supposed that all its lower-order singularities have already been shown to be harmless (i.e. real), so that the task is merely to show that its leading singularity is similarly harmless. This is achieved by showing that, because of the existence of the real region R_0 in which no Feynman graph can be singular, some part of the leading singularity surface is non-singular. But, by the discussion of § 2.1, a non-singular part is divided from singular parts usually only by the cuts attached to the lower-order singularities. If the latter have already been shown to be harmless, this enables one to show that the leading singularity is similarly harmless. We now briefly sketch the argument for this; for fuller accounts the reader is referred to the original papers. The other means of division of non-singular parts from singular parts we discuss in § 2.8.

Foundations of the induction procedure

The cuts attached to lower-order singularities may divide singular parts of the leading singularity surface Σ from non-singular parts, but the precise positions of these cuts are not determined. We may choose how we draw them; only their ends, the lower-order singularities, are fixed for us. So it is evidently important to study the intersections of Σ with the lower-order surfaces. We are only interested in those intersections that correspond to the same values of the α on the two surfaces; only at such points (sometimes called effective intersections) can a pinching configuration of the singularity $D = 0$ in α-space fall off the edge of the hypercontour. Further, we need only consider intersections with lower-order surfaces that are themselves singular on the physical sheet, for if Σ is to have different singularity behaviour on either side of the intersection, the lower-order surface must be singular in order to cause that difference. In the following discussion we call intersections that satisfy both the above conditions *critical intersections*.

To set the induction procedure going, we must first consider the simplest graphs, those whose Feynman integrals depend only on one of the variables, s, t, u. These graphs are those having the structure of any of the diagrams of Fig. 2.5.2. Of these, the normal thresholds are inevitably singular on the physical sheet, but we have seen in §2.5 that, if the values of the masses are suitably chosen, these are the only singularities that these graphs produce on the physical sheet. Such a situation occurs in the equal-mass case, or more generally if the external masses are not too large compared with the internal masses. We shall suppose that this is the case here.

The normal thresholds have the crucial property that their critical intersection with higher-order surfaces is at infinity, that is they are asymptotic to the higher-order curves where they have critical intersection with them. To see this we consider a graph having the structure of Fig. 2.6.1. To obtain the leading surface Σ we must put all the internal momenta, both those exhibited explicitly in the figure and those contained in the two blobs, on the mass shell, and must also take account of any geometrical constraints imposed on the momenta in either blob by the Landau conditions

$$\sum_{\substack{\text{each} \\ \text{loop}}} \alpha_i q_i = 0. \tag{2.6.1}$$

To obtain the critical intersection of Σ with the normal threshold that is the leading singularity for the graph obtained from Fig. 2.6.1 by contracting the two shaded circles to points, we must in addition put all the α for the lines in these circles equal to zero. Consider now also the vertex graphs obtained from Fig. 2.6.1 by contracting one of these circles to a point. It is evident that the conditions for the internal momenta of the original graph to be such as to correspond to a critical intersection of Σ with the normal threshold, are more than need be

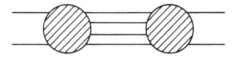

Fig. 2.6.1. A diagram showing the structure of the Feynman graphs that is relevant to the proof that critical intersections with normal thresholds are at infinity.

satisfied for the leading singularity of the vertex graph. In other words, the leading surface for the vertex graphs passes through the critical intersection in question. But both the normal threshold graph and the vertex graph correspond to Feynman integrals that involve only one of the variables s, t, u; for definiteness let it be s. So the normal threshold and the vertex singularity are just parallel straight lines $s = $ constant. These only meet at infinity.

General techniques

A general graph has normal thresholds corresponding to each of the three variables s, t, u. However at any point in the real plane at most two of these sets of cuts overlap. For example, in the equal-mass case the cuts occupy the three regions

$$s \geqslant 4m^2, \quad t \geqslant 4m^2, \quad u \geqslant 4m^2,$$

of the real plane. If we apply the linear constraint among the variables,

$$s + t + u = 4m^2,$$

these three regions appear in Fig. 2.6.2 as the regions to the sides of the lines $s = 4m^2$, $t = 4m^2$, $u = 4m^2$ marked by the arrows. The overlaps of two such regions are cross-hatched. In the central triangle is the region R_0.

The significance of the regions of overlap is that, as we saw in the examples of §§2.3 and 2.4, any piece of the real section of a Landau surface Σ lying in these regions may have different singularity behaviour depending on the choice of complex direction from which it is approached. We have also seen that there are at most two different types of behaviour; approaches from complex-conjugate directions must give essentially the same results. This means that in regions where there are normal-threshold cuts corresponding to only one variable all approaches must give the same behaviour, because the way in which the approach is made in the variables not having the cuts is immaterial.

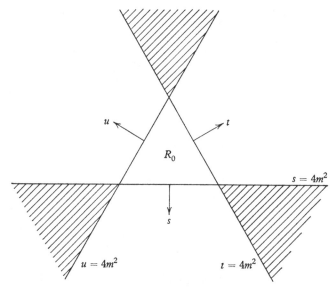

Fig. 2.6.2. The real region R_0, and the three regions (cross-hatched) where there are normal threshold branch cuts in two of the variables, s, t or u, for elastic scattering in an equal-mass theory.

The normal-threshold cuts divide a Landau surface into a number of sections. If, as is supposed in the induction procedure, there are no other cuts corresponding to lower-order singularities, the arguments of §2.2 require each of these sections to be either wholly singular or wholly non-singular. These sections are then linked together in chains, such that each chain passes through a region of complex space which is already known, from the existence of single-variable dispersion relations that we proved in section 2.5, to be free of singularity. So the whole chain must be non-singular.

We give an illustration of how we link together the sections to form chains. In Fig. 2.6.3a we draw a loop γ of the real section of some Landau surface Σ, lying in the region where the s and t normal-thresholds cuts overlap. As in §§ 2.3 and 2.4, the searchline method shows that from γ sprout four sections $\Sigma_A, \Sigma_B, \Sigma_C, \Sigma_D$ of complex surface. On Σ_A $\mathrm{Im}\, s/\mathrm{Im}\, t < 0,$

while on Σ_B $\mathrm{Im}\, s/\mathrm{Im}\, t > 0,$

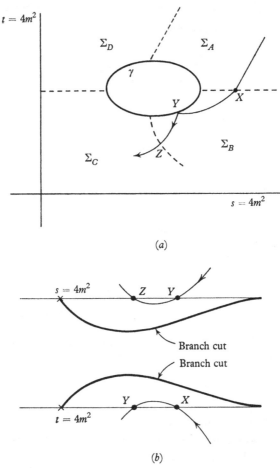

Fig. 2.6.3. (a) A real curve γ that is the real section of a Landau surface Σ which is divided into parts Σ_A, Σ_B, Σ_C, Σ_D by the normal threshold branch cuts. (b) The distortions of normal threshold branch cuts that are used in tracing a path on Σ that links Σ_A to Σ_C. The real points X, Y, Z are also shown in the drawing (a) above.

so that Σ_A and Σ_B are separated by the cut running along $\mathrm{Im}\, t = 0$. Similarly, Σ_B, Σ_C are separated by the s-cut, Σ_C and Σ_D by the t-cut and Σ_D, Σ_A by the s-cut. To link together Σ_A and Σ_C we must join them by a path that lies entirely in Σ and does not pass through any cut. To achieve this, we must temporarily push these cuts out of the way, which we are allowed to do because only the end-points of the cuts, the branch points, are rigidly fixed by the theory. (Alternatively, instead of pushing back the cuts, we may think of the path as temporarily passing on to another Riemann sheet). So we start at some point on Σ_A with, for definiteness,

$$\mathrm{Im}\,s > 0, \quad \mathrm{Im}\,t < 0.$$

We attempt to cross from Σ_A to Σ_B, but when we get to the point X on the boundary between the two, we have reached the cut $\mathrm{Im}\,t = 0$. We push this cut back, and so can move on to Σ_B, where now we have

$$\mathrm{Im}\,s > 0, \quad \mathrm{Im}\,t > 0.$$

We now come down on to γ at the real point Y and cross through into the region

$$\mathrm{Im}\,s < 0. \quad \mathrm{Im}\,t < 0$$

of Σ_B, pushing back the s-cut to allow this. We then cross over to Σ_C, going through the point Z at which $\mathrm{Im}\,s = 0$. The projections of our path on the complex s- and t-planes are shown in Fig. 2.6.3b. Provided we have not encircled any unwanted lower-order singularity we have now successfully linked Σ_A, Σ_C and may now restore the cuts to the real axes.

There is evidently a certain amount of freedom in the choice of the precise shape of path, and in fact the only lower-order singularities that we cannot avoid encircling are those that touch γ. In this case the path cannot avoid going on to the wrong Riemann sheet of that singularity and will not end up back on the physical sheet. This is why it is crucial that the critical intersections with the normal thresholds are at infinity, where they are harmless.

There are two difficulties with the above procedure. One is discussed in § 2.8. The other concerns what have been called *virtual singularities*. These are real arcs lying in crossed cuts, such as we encountered in §§ 2.3 and 2.4, that are singular from one method of approach, though not from those directions that would require the complex surfaces attached to them to be singular. Although these are harmless in that they are not associated with complex singularities, special arguments must be involved to ensure that they do not block the linking of the sections into chains. For these, and some other details, the reader is referred to the original papers (Eden, 1961; Landshoff *et al.* 1961).

The method of analytic completion

The foregoing discussion is based on linking different parts of the Landau surface Σ without crossing branch cuts that would lead off the physical sheet. With this linkage if a part of Σ is known to be non-singular on the physical sheet, one can deduce that any other part to which it is linked in this manner is also non-singular.

An alternative approach (Eden, 1961) is to establish a part of the physical sheet that is non-singular and then use the methods of 'analytic completion' (Wightman, 1960) to sweep out a tube that is free from singularities. We briefly review this method here.

If for some value of t the function $A(s,t)$ can be proved to be analytic within a closed curve C in the s-plane, then from Cauchy's theorem

$$A(s,t) = \frac{1}{2\pi i} \int_C \frac{A(s',t)\,ds'}{s'-s}. \tag{2.6.2}$$

If the contour C is displaced by varying t, the equation (2.6.2) can be used to define $A(s,t)$ for all points s within the tube swept by C during the displacement of C, *provided that* during the displacement no singularity of $A(s,t)$ in the integrand crosses the curve C. This has two consequences that are of great value in discussing domains of analyticity:

(A) we need consider only singularities on the boundary of C during such a displacement. If we can establish that no singularity meets the boundary during the displacement, then analytic completion tells us that there can be no singularity inside the boundary (within the tube).

(B) because no singularities can appear inside the closed curve C during such a displacement, there can be no singularity that would lead to a 'horn' of singularities projecting into the closed curve C. A horn of singularities is illustrated in Fig. 2.6.4. If the shaded region were singular but its surroundings were not singularities of $F(s,t)$, then a displacement from C to C' would not meet any singularities. Formula (2.6.2) provides a method of continuing analytically within C', hence the shaded projection cannot be singular. The fact that such projections cannot be singular is called the 'disc theorem'.

The method of analytic completion can be applied to a scattering amplitude using both the features A and B noted above. We will consider it for equal masses. We begin from our knowledge of analyticity of $A(s,t)$ in the upper half s-plane when t is real and in $(-4m^2,$

$4m^2$). The closed curve C used in equation (2.6.2) can be the large semicircle C in the upper half s-plane, illustrated in Fig. 2.6.5.

Keeping Re (t) in $(0 < t < 4m^2)$ we can now displace C to Im $(t) = i\epsilon'$. No singularities will be encountered by C since for t in $(0, 4m^2)$ the only

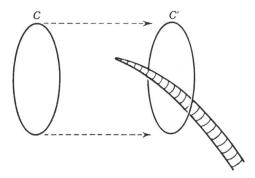

Fig. 2.6.4. A horn of singularities (shaded) that can be removed through analytic completion by displacing the contour of integration C of equation (2.6.2) to a new contour C'.

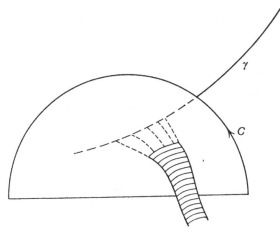

Fig. 2.6.5. A real Landau curve γ, showing that if it were singular in a certain limit its attached singular surface would be a horn of singularities penetrating the contour C, and could therefore be removed.

singularities near C are: (i) normal thresholds in s; $s = 4m^2$ for example; (ii) normal thresholds in u; $u = 4m^2$ for example.

A normal threshold in s is avoided by the line Im $(s) = i\epsilon$ (which is the nearest boundary of C), independently of the value of t. Similarly, a normal threshold in u is avoided by

$$\text{Im} (u) \equiv \text{Im} (4m^2 - s - t) = -i(\epsilon + \epsilon'),$$

provided ϵ' is positive. More generally in considering ϵ' negative it must be required that $|\epsilon'| < \epsilon$, which does not seem to be a serious restriction in this context.

The circular part of C is assumed to be of very large radius; it can therefore only meet a singularity that moves in from the point at infinity in the z-plane. Such behaviour cannot arise from a normal threshold so it is not relevant at this stage.

We next increase Re (t) to Re $(t) > 4m^2$ with the contour C having its straight part along Im $(s) = i\epsilon$, and with Im $(t) = i\epsilon'$. The normal threshold at $t = 4m^2$ is avoided by the imaginary part $i\epsilon'$ of t. If the only singularities were normal thresholds, the displacement could continue with Re $(t) \to \infty$. Similar displacements with Im $(t) = -i\epsilon'$, and also displacements with Re $(u) \to \pm\infty$, would then establish that $A(s,t)$ is analytic in the product of complex planes cut along the real s-, t-, and u-axes. However there are, in addition to normal thresholds, also curves of singularities in the real (s,t)-plane, which we must consider.

It can be established that certain general classes of Landau curves in the real (s,t)-plane do not block the process of analytic completion. But, as we shall see in § 2.8, these classes of curve do not include curves having acnodes or cusps. We illustrate the method by an example.

Since, in the displacement considered above, s and t have imaginary parts $i\epsilon$, $i\epsilon'$ of the same sign, the continued displacement of C along Im $(t) = i\epsilon'$ as Re $(t) \to \infty$, can be blocked only if there is a real singular curve γ having positive slope m, to which is attached a singular surface Σ having

$$\frac{\text{Im}\,(t)}{\text{Im}\,(s)} = m > 0. \tag{2.6.3}$$

If the second derivative also were positive, the surface Σ would extend downwards in the complex (s,t) space as indicated in Fig. 2.6.5. Thus if Σ were singular on the physical sheet, a horn of singularities would extend into C, contradicting the result (B) quoted above. Hence Σ cannot be singular on the physical sheet. Therefore if γ is singular on the boundary of the physical sheet (the product of cut planes), it must be singular only in the limit

$$\text{Im}\,s = 0 + i\epsilon, \quad \text{Im}\,t = 0 - i\epsilon', \tag{2.6.4}$$

taken from opposite half-planes. Then the singular part of Σ is reached only by going through the branch cuts along the real s-, t-axes; since (2.6.3) means that Σ lies in corresponding half-planes, but it is singular only when approached in the limit (2.6.4).

The only type of surface of singularity that can block the analytic completion method using displacement of C, is one that presents a concave surface to the approach of C. Thus the surface Σ must come out of the real (s, t)-plane through γ so that it slopes away from C. Then the boundary curve C strikes the singular surface and further displacement of C is blocked. In practice a surface of this awkward type can come out from γ at a cusp, or at an acnode, as we shall see in §2.8. Thus analytic completion suffers the same limitations as the methods described earlier in this section. However, both are useful in establishing that certain regions are free from singularities. The linkage of singular parts of the surface Σ is essential to both methods, but using analytic completion this linkage need be studied only in the neighbourhood of the real (s, t)-plane where the surface Σ is determined by the form of its real section, namely the curve γ.

2.7   Continuation in the external masses

If it can be proved that, for a certain set of values of the masses, there are no complex singularities in the scattering amplitude, we may then study what happens when the external masses are increased. As in the example of §2.3, the only way in which a complex section of a Landau surface can become singular is for the whole section to pass through a cut (by complex section we mean an area of the Landau surface with s, t, or both complex). Hence as the external masses are increased the first complex section to become singular will necessarily do so by first shrinking to zero. This is because before that happens there are no complex cuts, so that to become singular the section must go through an existing real cut, which it can only do by shrinking down to zero.

The square graph

We illustrate this process by considering the square graph. In §2.4 we found that, if the values of the masses are chosen such that the leading singularities corresponding to the triangle contractions are not singular, the square graph produces no complex singularities. The condition for this was that each y in (2.4.15) be less than one in modulus, and that the sum of any two of the corresponding θ, defined in (2.3.14), be less than π.

Suppose we now increase the external masses so that the sum of a pair of θ becomes greater than π. Then one of the triangle graphs becomes singular, say Σ_2 in Fig. 2.4.4, by moving up to the normal threshold Σ_{24} and away again. (To see what happens in detail we should have to avoid Σ_2 passing through Σ_{24} by temporarily giving the external masses a small imaginary part, as in §2.3. We should then

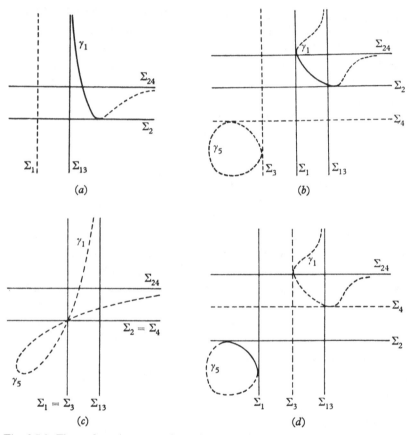

Fig. 2.7.1. The real sections γ_1 and γ_5 of the surface Σ, as the external masses are increased for the square diagram. For diagrams (a), (b), (c) there are no complex singularities on the physical sheet, for (d) there are complex singularities arising from the part of Σ joining γ_1 and γ_5.

find that Σ_2 passed *round* Σ_{24}, so that it passed through the cut attached to it. So the singularity has emerged from the neighbouring Riemann sheet of Σ_{24}.) As Σ_2 moves away from Σ_{24} it pulls the arc γ_1 with it, so that the picture now is as in Fig. 2.7.1a. In that figure parts of the curve corresponding to positive α are drawn in solid line, other parts in broken line. As usual these parts are identified by continuity arguments: on Σ_{13} we have

$$\alpha_1 = \alpha_3 = 0. \quad \alpha_2, \alpha_4 > 0.$$

On Σ_2, since it is now a singularity,

$$\alpha_2 = 0, \quad \alpha_1, \alpha_3, \alpha_4 > 0;$$

so on the part of γ_1 between its contact with Σ_2 and its asymptotic intersection with Σ_{13} all the α are positive. On the other hand, the other part of γ_1 has $\alpha_2 < 0$, since α_2 passes through zero at the contact with Σ_2.

We know, from the analysis of §2.4:

(i) That the boundary of R (the part of the real plane in which the Feynman integral is defined with undistorted hypercontour) is given by pieces of curve with positive α.

(ii) That for a piece of real curve in this region of the real (s, t)-plane to be singular from the limit

$$\{\operatorname{Im}(s)/\operatorname{Im}(t)\} > 0,$$

it must have positive α.

So in the situation of Fig. 2.7.1 a the region R is bounded by the lines Σ_{13} and Σ_2; it extends down to $s = -\infty$, $t = -\infty$. Also, the part of γ_1 between the contact with Σ_2 and the asymptotic intersection with Σ_{24} is singular in the limit with

$$\{\operatorname{Im}(s)/\operatorname{Im}(t)\} < 0,$$

but the pieces of complex surface attached to it lie in the regions having

$$\{\operatorname{Im}(s)/\operatorname{Im}(t)\} > 0.$$

Hence there is still no complex singularity. Indeed a complex singularity cannot have appeared, because no part of complex surface has shrunk to zero in the transition from the situation of Fig. 2.4.4 to that of Fig. 2.7.1 a. That this is so is most easily seen by noting that each piece of complex surface attached to γ_1 is also attached to one of the other real arcs $\gamma_2, \gamma_3, \gamma_4, \gamma_5$ of Fig. 2.4.4 (not drawn in Fig. 2.7.1 a), and none of these arcs has undergone any fundamental change in shape that would allow a piece of surface attached to it to degenerate.

If we now further increase the external masses, but in such a way that the sum of the four θ's remains less than 2π, either Σ_1 or Σ_3 will sooner or later move up to Σ_{13} and away again, becoming singular in the process and pulling γ_1 away with it. For definiteness let it be Σ_1; the situation is then as drawn in Fig. 2.7.1 b. Again no piece of complex surface has shrunk to zero, so there are still no complex singularities; each part of γ_1 is singular when approached from the limits opposite to those corresponding to the regions occupied by the attached pieces of complex surface. The region R is now bounded by Σ_1 and Σ_2.

If now the external masses are increased further, when the sum of the four θ's is equal to 2π the picture is as in Fig. 2.7.1c. The lines Σ_1 and Σ_3 have coalesced, also Σ_2 and Σ_4, and the real curve γ has degenerated. The two complex conjugate pieces of surface joining γ_1 and γ_5 have shrunk to zero. If the masses are further increased, so that the sum of the θ's becomes greater than 2π, these pieces re-emerge and so could be singular.

In fact they are singular. This is because, as indicated in Fig. 2.7.1d, the relevant piece of γ_5 now corresponds to positive α. So this part, together with parts of Σ_2 and Σ_1, provides the boundary of R and must be singular from any direction of approach. Hence the pieces of complex surface attached to it must be singular.

General case

The development of the crunode or double point indicated in Fig. 2.7.1c signals the onset of complex singularities on the physical sheet as the external masses are increased. The crunode and the subsequent complex singularities are closely related to the movement of the anomalous thresholds. This mechanism for the appearance of complex singularities on the physical sheet was the first and simplest to be discovered (Mandelstam, 1959). It is possible to show that, as the external masses are increased, the square graph is the first graph to acquire complex singularities in this manner (Eden, Landshoff, Polkinghorne & Taylor, 1961a). Any lines inserted into the graph delay the onset of these complex singularities. Unfortunately, as we shall see in §2.8, there are other less tractable mechanisms by which complex singularities may appear, and it is not yet known whether these occur first in the simpler graphs.

The first point to note for a general graph is that the development of complex singularities by the above mechanism must also be associated with the occurrence of a double point at the intersection of two (perpendicular) anomalous thresholds in the real (s,t)-plane. If a portion of a Landau curve is to have attached to it a surface corresponding to singularities of the amplitude on the physical sheet (as in Fig. 2.7.1d), the portion of curve must be singular when approached from the same side as the attached surface. Conversely, before the singularities on the physical sheet appear (that is with smaller external masses), the portion of the Landau curve must be singular only when approached from a direction opposite to that occupied by the attached surface. Thus in Fig. 2.7.1b the solid-line portion of γ_1, is singular only in the limit $(\mathrm{Im}\,s/\mathrm{Im}\,t) > 0$. The broken-line portion of γ_1 is singular only from the limit $(\mathrm{Im}\,s/\mathrm{Im}\,t) < 0$. The crucial point here is that the change from singularity in one type of limit to singularity in the other type occurs only at the point of tangency of γ to the anomalous thresholds. Since we require the limit that is relevant for a portion of γ to switch from one type to the other when it shrinks to a crunode and expands again, it is evident that this can occur only if the points of tangency to the anomalous thresholds coincide at the crunode when it forms. Then, and only then, can the attached surface switch from being singular on the unphysical side of the real branch cut (the $s + i\epsilon$, $t - i\epsilon$, side) to being singular on the physical side (the $s + i\epsilon$, $t + i\epsilon$, or the $s - i\epsilon$, $t - i\epsilon$, side in this example).

Thus we must, for the general graph, study the situation in which a crunode coincides with the intersection of an anomalous threshold in s and an anomalous threshold in t. We will indicate briefly how it can be seen that the insertion of lines into the square graph will delay the onset of complex singularities when the external masses are increased. Our discussion here is limited to the above mechanism involving a crunode and anomalous thresholds. The method is based on dual diagrams, which were introduced in § 2.3.

We consider first the occurrence of anomalous thresholds from the vertex parts formed by reducing a scattering diagram. A vertex dual diagram must be drawn in a plane in order to give an anomalous threshold. (It will be recalled that dual diagrams may be drawn in complex Euclidean space.) It follows that for fixed masses the anomalous threshold (if any) from the lowest order vertex occurs at a smaller value of the appropriate variables (s, t or u) than any such threshold from a higher order vertex part. This result comes from the fact that for a physical-sheet anomalous threshold the dual diagram can be drawn in real Euclidean space (Taylor, 1960). If the lowest order diagram does not give an anomalous threshold, nor does any other.

The dual diagram for a general scattering diagram is drawn in three dimensions. The dual of the square diagram is shown in Fig. 2.4.2. For a higher-order dual diagram the tetrahedral framework $ABCD$ remains the same, since it refers only to the external variables $\sqrt{s}$, $\sqrt{t}$ and the external masses, but the internal structure is more complicated.

For a critical double point on the Landau curve, as in Fig. 2.7.1c, there must be a simultaneous coincidence of the two perpendicular anomalous thresholds and the Landau curve. This requires that two parts of the corresponding dual diagram must be simultaneously coplanar and real Euclidean since the crunode lies on the boundary of R_0. Since for coplanar dual diagrams the lowest anomalous threshold comes from the simplest such diagram it can be seen that the required simultaneous coplanarity occurs first for the dual of the square diagram. Thus increasing external masses give rise to a crunode and to complex singularities for the square diagram before they arise from more complicated diagrams.

2.8   The acnode graph

There are two matters that prevent the techniques of §2.6 yielding a complete proof that, for suitable values of the masses, each graph that contributes to a scattering amplitude is free of complex singularities. Both these points are illustrated by a particular example, that of the Feynman graph of Fig. 2.8.1a (Eden, et al. 1961b). Other examples have since been discovered by Olive & Taylor (1962) and Islam (1963).

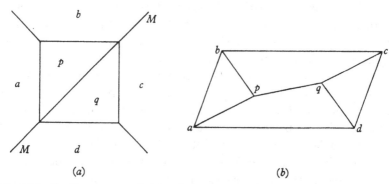

(a)   (b)

Fig. 2.8.1. The Feynman graph (a) and the corresponding dual diagram (b), that can lead to acnodes and cusps on the Landau curve.

The form of the Landau curve

The dual diagram for the graph is drawn in Fig. 2.8.1b. The Landau equations require each line to be on the mass shell for the leading singularity, and also the conditions

$$\sum_{\substack{\text{each loop} \\ \text{of the} \\ \text{graph}}} \alpha_i q_i = 0$$

require that the three lines connected to p in the dual diagram lie in a plane, and also the three lines connected to q. To obtain the equation of the leading singularity surface is now a matter of geometry; one has to find the constraint that the figure imposes between the lengths $\sqrt{s}$ and $\sqrt{t}$ of the diagonals ac, bd. The calculations are simplified considerably if a symmetrical set of values is assigned to the masses. Accordingly, we set each mass equal to unity except for the two external masses bc and ad which we take to be M. Then the solution of the Landau equations having the symmetry

$$\alpha_{ap} = \alpha_{cq}, \quad \alpha_{bp} = \alpha_{dq}$$

is readily found to be

$$\left. \begin{array}{l} s = 5 + 4\cos\phi + 2(2 - \tfrac{1}{2}M^2 + \cos\theta + \cos\phi)\sin\phi/\sin\theta, \\ t = 5 + 4\cos\theta + 2(2 - \tfrac{1}{2}M^2 + \cos\phi + \cos\theta)\sin\theta/\sin\phi, \end{array} \right\} \quad (2.8.1)$$

with $\theta + \phi = \tfrac{1}{3}\pi$. θ and ϕ are in fact the angles between pq and ap, bp produced. These angles are also equal to the angles between pq and cq, dq produced. There are other portions of curve having different symmetry; these we do not investigate.

The Feynman parameters are given by

$$\frac{\alpha_{ap}}{\sin\phi} = \frac{\alpha_{bp}}{\sin\theta} = \frac{2\alpha_{pq}}{\sqrt{3}} = \frac{\alpha_{cq}}{\sin\phi} = \frac{\alpha_{dq}}{\sin\theta}, \quad (2.8.2)$$

so that they are all real and positive when

$$0 < \theta < \tfrac{1}{3}\pi, \quad 0 < \phi < \tfrac{1}{3}\pi.$$

The part of the real curve for which θ and ϕ satisfy these conditions is drawn for various values of M^2 in Fig. 2.8.2. As may be seen from equation (2.8.1), the curve is asymptotic to the normal threshold $s = 9m^2$ for $\phi = 0$; this normal threshold is the contraction

$$\alpha_{ap} = \alpha_{cq} = 0$$

in agreement with equation (2.8.2). Again, the value $\theta = 0$ yields asymptotic intersection with the normal threshold $t = 9m^2$, corresponding to the contraction

$$\alpha_{bp} = \alpha_{dq} = 0.$$

To obtain the form as M^2 increases we write

$$\theta = \tfrac{1}{6}\pi + i\eta, \quad \phi = \tfrac{1}{6}\pi - i\eta. \quad (2.8.3)$$

When $\qquad 4 + 2\sqrt{2} < M^2 < 4 + 5/\sqrt{3} = 6{\cdot}887,$

it is found that there are real non-zero values of η such that

$$\operatorname{Im} s = \operatorname{Im} t = 0.$$

Thus at $M^2 = 4 + 2\sqrt{2}$ coincident isolated real points appear and then separate as M^2 is increased, to give the situation of Fig. 2.8.2b. Such isolated real points are known to geometers as *acnodes*. As M^2 passes through the value $4 + 5/\sqrt{3}$ one of the acnodes meets the continuous

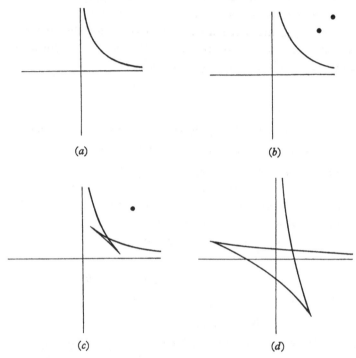

Fig. 2.8.2. The Landau curve for the 'acnode diagram' shown in Fig. 2.8.1a, showing the effect of increasing the external mass giving a variation from (a) through (b) and (c) to (d). The dots denote acnodes (isolated real points).

part of the real curve, and changes into a different type of node, a *crunode*. This is a point where a real arc crosses itself; we now have the situation of Fig. 2.8.2c. [We already encountered a crunode, in Fig. 2.7.1c, but that one only existed for special sets of values of the external masses. The present one is there for all $M^2 > 4 + 5/\sqrt{3}$.] The curve now also has a pair of real cusps; the position of these may be found from the conditions

$$\frac{ds}{d\theta} = \frac{dt}{d\theta} = 0. \qquad (2.8.4)$$

For $M^2 < 4 + 5/\sqrt{3}$ the equations (2.8.4) do have solutions, but for complex s, t: they correspond to complex cusps.

As M^2 increases further, the two cusps cross the normal thresholds; and for

$$7 > M^2 > 4\sqrt{3} = 6{\cdot}928,$$

part of the arc lies outside the normal-threshold cuts, as is drawn in Fig. 2.8.2d. Above $M^2 = 7$ the vertex contractions produce new singularities and the character of the curve changes in a way we need not consider.

The complex surfaces attached to the real arc are of complicated shape. We consider intersections with the particular set of searchlines

$$s + t = \mu, \quad \mu \text{ real.} \tag{2.8.5}$$

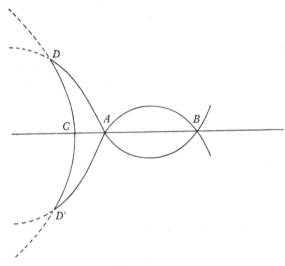

Fig. 2.8.3. The paths in the complex s-plane on the Landau surface, that are traced out by the searchline (2.8.5) when there are two acnodes A and B. This is the situation shown in Fig. 2.8.2b.

The paths traced out in the complex s-plane as μ is varied are shown for the configuration of Fig. 2.8.2b, in Fig. 2.8.3. The acnodes are labelled A, B, the midpoint of the real arc is denoted by C, and D, D' are complex cusps. The parts of the paths in Fig. 2.8.3, drawn in solid line correspond to real η and those in broken line to complex η.

Singularity properties

When the real arc extends outside the normal-threshold cuts, as in Fig. 2.8.2d, since it corresponds to positive α it forms part of the boundary of the region R for the graph, in which the Feynman integral is defined for undistorted hypercontour of integration. Hence that part is singular from any direction of approach, and therefore also the two complex-conjugate pieces of complex surface attached to it are singular on the physical sheet. Since the intersection of the arc with the normal thresholds is not an effective intersection (the values of the α are different on the arc from on the normal threshold at the intersection) the whole part of the arc between the cusps must have the same singularity property, and therefore also the whole of the attached pieces of complex surface. At their other ends these pieces of complex surface are attached to the acnode, as may be seen by considering searchlines.

At the cusps the second mechanism for division of a surface into singular and non-singular portions is involved; so that the complex surfaces attached to the rest of the real arc do not have to be singular. This is the mechanism discussed in connection with (2.1.13) and subsequent equations, or rather its analogue for multiple integrals. That this occurs at a cusp may readily be seen. The denominator function D takes the form

$$D = f(\alpha)\,s + g(\alpha)\,t + K. \tag{2.8.6}$$

On the leading Landau curve each

$$\frac{\partial D}{\partial \alpha_i} = 0. \tag{2.8.7}$$

But, since D is homogeneous in the α, so is $\partial D/\partial \alpha_i$. Therefore, by Euler's theorem,

$$\sum_j \frac{\partial^2 D}{\partial \alpha_i \partial \alpha_j} \alpha_j \propto \frac{\partial D}{\partial \alpha_i}.$$

So (2.8.7) implies that the $N \times N$ matrix $\partial^2 D/\partial \alpha_i \partial \alpha_j$ has zero determinant: the Landau curve is characterized by this matrix having rank $(N-1)$ instead of N. The matrix determines the form of the surface $D = 0$ in the neighbourhood of the critical point in α-space. When its rank is $(N-1)$ this form is locally cone-like, and the possibility of a singularity of the Feynman integral arises from trapping of the hypercontour of integration between the two halves of the cone. Now if we vary the parameters α so that we move to another point of the Landau curve, from (2.8.6) and (2.8.7) the displacement (ds, dt) along the curve is given by

$$0 = d\left(\frac{\partial D}{\partial \alpha_i}\right) = \frac{\partial f}{\partial \alpha_i}\,ds + \frac{\partial g}{\partial \alpha_i}\,dt + \sum_j \frac{\partial^2 D}{\partial \alpha_i \partial \alpha_j}\,d\alpha_j. \tag{2.8.8}$$

But if we are at a cusp a small variation of the α does not change s, t, as may be seen from (2.8.4). So then

$$0 = \sum_j \frac{\partial^2 D}{\partial \alpha_i \partial \alpha_j} d\alpha_j.$$

Since the $d\alpha$ are not proportional to the α (we recall that only the *ratios* of the α are involved in the Landau equations; their absolute magnitudes would come in only if we were to make use of the δ-function in the numerator of the Feynman integral (2.2.4)) this means that the matrix $\partial^2 D/\partial \alpha_i \partial \alpha_j$ is further reduced in rank, to $(N-2)$. So the cone has degenerated and the intersection of $D = 0$ with the hypercontour involves more points than are necessary for the trapping. This is the multi-dimensional analogue of the situation illustrated in Fig. 2.1.6.

If we now reduce M^2, so passing from Fig. 2.8.2d to Fig. 2.8.2c, the pieces of complex surface joining the part of the real arc between the cusps to the acnode must remain singular, because it has not shrunk to zero. If we reduce M^2 still further, so as to arrive at Fig. 2.8.2b, these pieces of complex surface have broken away from the real arc and now just join the two acnodes; they must still be singular.

So now we have complex singularities even though there are no real cusps. The analysis of § 2.6 has broken down because of the acnodes, which do not allow us to link together the sections of complex surface in chains. It was originally thought that acnodes, being degenerate points of curves, would occur only for accidentally chosen sets of values of the masses, so that if the masses were varied slightly the acnode would become a real closed loop and the arguments used for Fig. 2.6.3 would apply. However, the present example nullifies this hope.

It is easy to see that the acnodes persist if the masses are changed slightly so as to give a more general diagram. The simplest direct way is to note that the derivative ds/dt in the Landau surface at the acnode is complex, and that it remains complex when the masses are varied slightly. Hence the isolated real point on the Landau surface does not become a real curve since for a real curve the derivative would have to be real.

An alternative argument can be based on the fact that a complex part of the Landau surface is singular and is attached to the acnode. If the acnode were to become a small real closed curve under a small mass variation, all this curve would have to be singular in the same limit (from opposite half planes in s and t). This would imply the existence of another singular part of the Landau surface attached to the other side of the closed curve. From this one can readily derive a contradiction by varying the masses to a situation where the arguments used for Fig. 2.6.3 would show that there is no complex singularity on the physical sheet.

What goes wrong with the analysis associated with Fig. 2.6.3 when there are acnodes is that here the two complex-conjugate portions of surface are not connected by a path through the real curve such that the α change continuously on this path. Here the arcs BAD and BAD' in Fig. 2.8.3 share common values of the α neither at A nor at B.

To sum up, in this example complex singularities are absent only for $M^2 < 4 + 2\sqrt{2}$. The general analysis of §2.6 cannot deal with acnodes and cusps, so it is not known whether or not *all* Feynman graphs are similarly free of complex singularities below this value of M^2.

The general problem of determining whether acnodes or cusps occur on the boundary of the physical sheet so that they are associated with complex singularities has not yet been solved. It cannot be solved by any of the majorisation methods such as we have presented in §2.5 because acnodes are always associated with a distortion of the α hypercontour in the corresponding Feynman integral. This follows from the fact that the derivative in the Landau surface must be complex at an acnode (otherwise it would not be an isolated real point). But the derivative ds/dt is the ratio of two functions of α that would be real unless the α variables themselves are complex. The study of Feynman diagrams for undistorted contours used majorisation techniques for which the real positive character of the Feynman parameters is vital. Their possible extension to complex contours is an important unsolved problem.

2.9   Discontinuities and generalised unitarity

In §2.1 we explained how to find the discontinuity associated with a branch-point of a function given by a simple integral. A Feynman integral is a multiple integral and so the corresponding analysis is considerably more complicated. However, the results are of simple appearance (Cutkosky, 1960; see also Fowler, 1962), and, when the discontinuities for all the graphs containing a given singularity are added together, take a form that is independent of the structure of the individual terms of the perturbation expansion from which they are derived. This encourages the hope that they are valid independently of perturbation theory, which itself is not valid in a numerical sense.

The discontinuity formula

Consider a general Feynman integral in the representation (2.2.1):

$$I(z) = \int \frac{d^4k_1 \, d^4k_2 \dots d^4k_l}{\prod\limits_{i=1}^{N} (q_i^2 - m_i^2)}, \qquad (2.9.1)$$

and suppose it is required to find the discontinuity associated with the singularity corresponding to $(N-r)$ contractions. If we label the q suitably the relevant singularity conditions are

$$q_i^2 = m_i^2 \quad (i = 1, 2, \dots, r) \qquad (2.9.2)$$

$$\alpha_i = 0 \quad (i = r+1, r+2, \dots, N), \qquad (2.9.3)$$

$$\sum_{\substack{\text{each} \\ \text{loop}}} \alpha_i q_i = 0. \qquad (2.9.4)$$

In the integral (2.9.1) there are $4l$ integration variables. Make a transformation of integration variables so that the integration is over the r variables $q_1^2, q_2^2, \dots, q_r^2$, together with $4l - r$ other, unspecified, variables ξ. (We consider later what happens if it should be that $4l < r$.) The transformation of variables will, of course, involve a Jacobian J, so (2.9.1) now reads

$$I(z) = \int_{a_1}^{b_1} dq_1^2 \int_{a_2}^{b_2} dq_2^2 \dots \int_{a_r}^{b_r} dq_r^2 \int \frac{\Pi d\xi}{J \prod\limits_{i=1}^{N} (q_i^2 - m_i^2)}. \qquad (2.9.5)$$

The limits (a_j, b_j) for the q_j^2 integration are the extrema of q_j^2 with respect to all values of the k when q_i^2, for $i < j$, are fixed. These we may find by introducing Lagrangian multipliers β and solving the equations

$$\sum_{\substack{\text{each} \\ \text{loop} \\ i \leqslant j}} \beta_i q_i = 0. \qquad (2.9.6)$$

Suppose now we write (2.9.5) as

$$I(z) = \int_{a_1}^{b_1} dq_1^2 \frac{I_1(q_1^2, z)}{q_1^2 - m_1^2}. \qquad (2.9.7)$$

The singularity $z = z_0$ of I that we are investigating depends in position on the value of m_1^2, but neither $I_1(q_1^2)$ nor a_1, b_1 explicitly contains m_1^2. Hence the singularity must arise in the integration in (2.9.7) as a result of the contour of integration being trapped between the fixed singularity produced by the denominator of the integrand and a singularity $\tilde{q}_1^2$ of I_1 that approaches m_1^2 when $z \to z_0$. So for z near z_0 the situation in the q_1^2 plane is as in Fig. 2.9.1a. We may split the contour of integration in this figure into two parts, as in Fig. 2.9.1b; then only the closed part of the contour will be trapped as $z \to z_0$ and so the integral over the open part will not contain the singularity. So the latter part will give zero discontinuity and we may forget it. The integral over the closed part is simple to evaluate; it is

$$- 2\pi i\, I_1(m_1^2, z). \qquad (2.9.8)$$

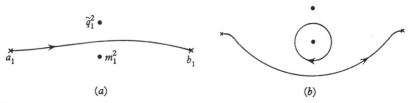

$$(a) \qquad\qquad\qquad (b)$$

Fig. 2.9.1. The method of splitting a contour of integration into two parts that is used in obtaining the discontinuity formula.

But we may write
$$I_1 = \int_{a_2}^{b_2} dq_2^2 \frac{I_2(q_2^2, z)}{q_2^2 - m_2^2},$$

and so repeat the argument. So eventually we find that the part of I that matters is
$$\int_{a_r}^{b_r} dq_r^2 \frac{I_r(q_r^2, z)}{q_r^2 - m_r^2}. \qquad (2.9.9)$$

In the q_r^2 integration the singularity of I comes about not from a trapping of the integration contour between two singularities, but because, when $z \to z_0$, one of the end-points of the integration moves towards the singularity $q_r^2 = m_r^2$ arising from the denominator of the integrand. This is because the end-points (a_j, b_j) are given by (2.9.6), which for $j = r$ is equivalent to (2.9.3) together with (2.9.4). We may now find the discontinuity of I by applying the discussion of Fig. 2.1.3. If we revert to the variables of integration k the result is

$$\operatorname{disc} I = (-2\pi i)^r \int \frac{(\Pi d^4 k)\, \delta^{(+)}(q_1^2 - m_1^2)\, \delta^{(+)}(q_2^2 - m_2^2) \dots \delta^{(+)}(q_r^2 - m_r^2)}{(q_{r+1}^2 - m_{r+1}^2) \dots (q_N^2 - m_N^2)}.$$
$$(2.9.10)$$

A word must be said about the meaning of the $\delta^{(+)}$-functions in equation (2.9.10). They can only be taken literally in the case of physical-region singularities, where, as we saw in §2.5, the critical values of the momenta q are real and physical. This will not usually be the case for singularities out of the physical region, and then the only reliable way to interpret (2.9.10) is to make an analytic continuation in some of the masses so that the singularities come into the physical region; then evaluate (2.9.10), and continue the result back to the desired values of the masses. This procedure is not simple in practice. It is because we were implicitly considering physical-region singularities that we drew m_1^2 below the contour in Fig. 2.9.1a—in the physical region the contour of integration is real and Feynman's $-i\epsilon$ depresses the masses into the lower-half plane. Otherwise m_1^2 and $\tilde{q}_1^2$ might be interchanged, and a difference of sign would result in (2.9.8).

Generalised unitarity

We know that a given singularity $z = z_0$ is shared by an infinite number of Feynman graphs; any graph that can be contracted suitably has that singularity. For example, the triangle singularity that we discussed in §2.3 is contained in all graphs of the structure of Fig. 2.9.2a, where the three blobs in the figure have any structure. The

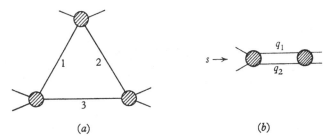

(a)   (b)

Fig. 2.9.2. (a) The general triangle singularity from this diagram has the discontinuity shown in equation (2.9.11). (b) A normal threshold singularity.

discontinuity, for each graph, associated with the triangle singularity takes the form (2.9.10), with $\delta^{(+)}$-functions for the three internal lines drawn explicitly in the figure and propagators $(q^2 - m^2)^{-1}$ for the others. If we sum over all graphs of this structure, we evidently obtain for the total discontinuity

$$(-2\pi i)^3 \int d^4k \, \delta^{(+)}(q_1^2 - m_1^2) \, \delta^{(+)}(q_2^2 - m_2^2) \, \delta^{(+)}(q_3^2 - m_3^2) A_1 A_2 A_3, \quad (2.9.11)$$

where A_1, A_2, A_3 are the sums of all the subgraphs contributing to the blobs, that is they are the complete scattering amplitudes represented by the blobs. So the result (2.9.11) takes a form independent of the perturbation theory from which it is derived, and we might expect to be able to derive it without using Feynman graphs. This we do in §4.11.

One point requires care in this argument. If we go round the triangle singularity with one of the variables, there is no necessity of simultaneously going round a singularity of one of the blobs. But consider the example of Fig. 2.9.2 b, and suppose it is desired to find the total discontinuity associated with the normal threshold corresponding to the two internal lines q_1, q_2 displayed explicitly in the figure. This normal threshold will also be contained in each of the two blobs, so if we go round it in the external variable s we must consider how the blobs change as well as taking account of the contours of integration associated with q_1^2, q_2^2 (Olive 1963 b). We know, from the unitarity equation (1.3.5) what the result is for the discontinuity: the structure of the blobs results in one of the scattering amplitudes occurring in the discontinuity formula being evaluated on a different Riemann sheet; in this particular case this usually involves the complex-conjugate of the physical amplitude, as in equation (1.3.5). These matters are investigated further in §4.6.

Simple examples

We give two examples of calculation of the discontinuity functions for simple graphs. First consider the triangle graph of Fig. 2.9.3 a.

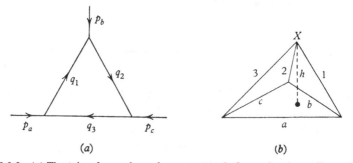

$$(a) \qquad\qquad\qquad (b)$$

Fig. 2.9.3. (*a*) The triangle graph used as an example for evaluating a discontinuity. (*b*) The momentum-space diagram for the triangle graph.

To get the discontinuity corresponding to the leading singularity, we have to evaluate (2.9.11), with A_1, A_2, A_3 set equal to the coupling constants for the vertices. We transform to the variables

$$q_1^2, \quad q_2^2, \quad q_3^2, \quad \phi,$$

where, in the momentum-space diagram of Fig. 2.9.3b, the angle ϕ measures the position of X on the circle of radius h in four-dimensional space on which X can lie when the lengths q_1^2, q_2^2, q_3^2 are given. We have

$$d^4k = h\,d\phi\,d^3k = \frac{h\,d\phi\,dq_1^2\,dq_2^2\,dq_3^2}{J}, \qquad (2.9.12)$$

where, since each q depends linearly on k, the Jacobian J is the three by three determinant $8\det \mathbf{q}_i$. But $\det \mathbf{q}_i$ is six times the volume of the tetrahedron in Fig. 2.9.3b, which is $\frac{1}{3}hA$, where A is the area of the base triangle. The integrations are trivial, so, apart from constant factors, the result is $1/\sqrt{K}$, where

$$K = p_a^4 + p_b^4 + p_c^4 - 2p_a^2 p_b^2 - 2p_a^2 p_c^2 - 2p_b^2 p_c^2. \qquad (2.9.13)$$

One reason for particular interest in the triangle singularity is that, according to (2.2.16) and (2.2.17), it is one of the few singularities that is an infinity rather than just a finite branch-point. So when it comes close to the physical region it might be expected to make its presence felt directly as a 'bump' in the amplitude (Landshoff & Treiman, 1962). The magnitude of this effect is estimated by using the discontinuity in a dispersion relation.

The other example we consider is that of Fig. 2.9.4a. This is an example where there are more internal momenta q than integration

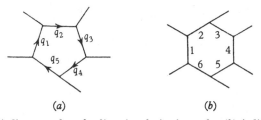

(a)   (b)

Fig. 2.9.4. (a) A diagram whose leading singularity is a pole. (b) A diagram in which there is no 'leading singularity' because of the constraints of four-dimensional space-time.

variables, so $4l < r$ when we are considering the leading singularity. But in fact the result (2.9.10) is still valid. If we accept it, for the leading singularity there are five δ-functions in the integral but only four integrations, so the result of the integration contains a δ-function as a factor. The singularity whose discontinuity is a δ-function is a pole, so we conclude that the leading singularity of the graph in Fig. 2.9.4a is a pole. This is corroborated by calculation of the discontinuity across the singularity corresponding to one contraction, say $\alpha_5 = 0$. For this singularity $4l = r$ and the derivation of (2.9.10) is valid. It gives an integral over four variables whose integrand contains four $\delta^{(+)}$-functions, making the integrations trivial, and the pole $(q_5^2 - m_5^2)^{-1}$. The $\delta^{(+)}$-functions, when integrated, require q_5^2 to be expressed in terms of the external momenta and the other internal masses. The pole is the leading singularity.

In conclusion, we note that if we were to apply (2.9.10) to try and find the discontinuity corresponding to the leading singularity of the hexagon graph of Fig. 2.9.4b, we should get an integral over four variables and containing six δ-functions. The reason for this non-sensical result is that the leading singularity does not in fact exist (Landshoff, 1960). In four-dimensional Lorentz space the dual diagram leads to *two* constraints on the external momenta instead of the usual one. But a singularity of a function of several complex variables must be given by one equation, not two simultaneous equations. If Lorentz space had five, or more, dimensions the singularity would exist.

2.10   Second-type singularities

Simple examples

We see from (2.9.13) that the discontinuity (2.9.11) associated with the triangle diagram is singular when

$$s_1^2 + s_2^2 + s_3^2 - 2s_2 s_3 - 2s_3 s_1 - 2s_1 s_2 = 0, \qquad (2.10.1)$$

where $s_i = p_i^2$ $(i = 1, 2, 3)$, p_i being an external four-momentum. This is not a singularity corresponding to any Landau curve associated with a dual diagram for it does not depend upon the internal masses. It is in fact the condition for the three external momenta of the triangle to lie along a line instead of in a plane. Since the discontinuity is just the difference of the Feynman integral on two of its Riemann sheets, the integral itself must also possess this singularity on at least one of the two sheets. This was first noticed by Cutkosky (1960) who gave the singularity the name 'non-Landauian', since it did not appear to be associated with the Landau equations. In fact, however, it turns out that this is an example of a wide class of singularities corresponding to rather special solutions of the Landau equations (Fairlie, Landshoff, Nuttall & Polkinghorne, 1962a, b). Members of this class are called *second type* singularities.

The reason that they do not appear among the dual diagram solutions of the Landau equations is very simply illustrated by considering the diagram of Fig. 2.10.1. The Feynman integral is divergent for a four-dimensional loop momentum k so we evaluate it with three-dimensional momenta. The dual diagram (first type) singularities are just the normal and pseudo thresholds

$$s = p^2 = (m_1 \pm m_2)^2. \tag{2.10.2}$$

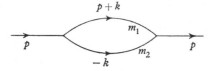

Fig. 2.10.1. This self-energy diagram gives rise to a second-type singularity at $s = 0$.

The discontinuity of the three-dimensional integral around the normal threshold is readily calculated by methods similar to the derivation of (2.9.11) and it is found to be proportional to

$$s^{-\frac{1}{2}}, \tag{2.10.3}$$

exhibiting a second-type singularity at $s = 0$.

Now the Landau equations for Fig. 2.10.1 are

$$(p+k)^2 - m_1^2 = 0, \quad k^2 - m_2^2 = 0, \tag{2.10.4}$$

$$\alpha_1(p+k) - \alpha_2 k = 0. \tag{2.10.5}$$

A solution of these equations can be given by drawing two spheres in momentum space, of radii m_1 and m_2, whose centres are joined by the vector p. Equation (2.10.5) then requires that the two spheres touch. If the distance between their centres is either the sum or the difference of their radii the two spheres touch at a finite point. This corresponds to the first-type solution (2.10.2). If, however, the two spheres are concentric (or more generally, in complex Euclidean space, if their centres are joined by a zero-length vector) then the spheres touch at infinity. Thus the reason that second-type singularities were at first overlooked lies in the fact that they correspond to pinch configurations in which some of the components of the loop momenta are infinite.

Figure 2.10.1 also serves to illustrate another interesting feature of second-type singularities. If the diagram is evaluated with a *two-dimensional* momentum k the second-type singularity is absent from the discontinuity and, as it turns out, from the integral itself. Thus the presence or absence of second-type singularities is dependent upon the dimension of space.

α-space analysis

The second-type singularities we have encountered are all given by equations of Gram-determinant type

$$\det p_i \cdot p_j = 0 \quad (i,j = 1, ..., n);$$ (2.10.6)

the two particular examples discussed having corresponded to $n = 2$, and $n = 1$ respectively. Equation (2.10.6) is the condition that there be a linear combination of the vectors $p_1 ... p_n$ equal to zero or, more generally, equal to a zero-length vector whose scalar products with $p_1, ..., p_n$ are zero.

If a diagram has E external lines there are $E - 1$ independent external momenta when account is taken of energy-momentum conservation. We might expect that there will be a second-type singularity when equation (2.10.6) is applied to these independent momenta. For $E \geqslant 6$ the condition is always satisfied, since Lorentz-space is four-dimensional and there must consequently be a linear relation between any five vectors in it. Since the whole of space cannot be singular the second-type singularity must be absent in this case. In fact, for a reason explained later, it is also absent for $E = 5$. However, the singularities are found for $E \leqslant 4$ and we proceed to give their explanation.

We use the notation of §1.5 augmented by the convention that matrix quantities which are also Lorentz vectors will have an arrow superscript to emphasise their status. The Landau equations include the relation

$$\mathbf{A}\vec{\mathbf{k}} = \mathbf{B}\vec{\mathbf{p}}.$$ (2.10.7)

If $\mathbf{A}$ is non-singular (2.10.7) will have a unique solution in terms of the p's which will exactly correspond to the dual diagram construction. Hence second-type solutions will have to correspond to $\mathbf{A}$ being singular, that is to the condition

$$C = \det \mathbf{A} = 0.$$ (2.10.8)

In this case $\mathbf{X} = \text{adj}\,\mathbf{A}$ can be written in terms of a column matrix $\mathbf{K}$,

$$\mathbf{X} = \mathbf{K}.\mathbf{K}^T,$$ (2.10.9)

where $$\mathbf{K}^T.\mathbf{A} = 0.$$ (2.10.10)

If the equations (2.10.7) are to have a solution with $C = 0$, (2.10.10) requires that

$$\mathbf{K}^T.\mathbf{B}\vec{\mathbf{p}} = 0,$$ (2.10.11)

in which case there are an infinite number of solutions for $\vec{k}$. The equation (2.10.11) just represents a linear relationship among the vectors p, which leads to the Gram-determinant equation. However, the most general way of satisfying the Gram-determinant relation involves a zero-length vector. To obtain this possibility we replace (2.10.7) by

$$\mathbf{A}\vec{k} = \mathbf{B}\vec{p} - \mathbf{\Lambda}\vec{\lambda}, \qquad (2.10.12)$$

where $\mathbf{\Lambda}$ is any column vector and $\vec{\lambda}$ is a Lorentz vector satisfying

$$\vec{\lambda}^2 = 0, \quad \vec{\lambda}.\vec{p} = 0 \text{ for all } \vec{p}. \qquad (2.10.13)$$

This leads in the case $C = 0$ to

$$\mathbf{K}^T.\mathbf{B}\vec{p} = \mathbf{K}^T.\mathbf{\Lambda}\vec{\lambda}, \qquad (2.10.14)$$

corresponding to the most general solution of the Gram-determinant condition. The status of the somewhat curious equation (2.10.12) will be made clear a little later in the argument.

Elementary manipulations now lead to the result that for $C = 0$ we can write

$$\frac{\partial D}{\partial \alpha_i} = \psi(\vec{p}, \vec{k}, \alpha)\frac{\partial C}{\partial \alpha_i}, \qquad (2.10.15)$$

provided that the solutions $\vec{k}$ of (2.10.12) are chosen to be linear combinations of the $\vec{p}$'s so that

$$\vec{\lambda}.\vec{k} = 0. \qquad (2.10.16)$$

Equation (2.10.15) shows that for $C = 0$ we can obtain a pinch configuration corresponding to

$$\frac{\partial D}{\partial \alpha_i} = 0 \quad (i = 1, 2, \ldots), \qquad (2.10.17)$$

by choosing the $\vec{k}$ as defined above and satisfying the single further condition

$$\psi = 0. \qquad (2.10.18)$$

There are many such solutions possible.

Momentum space analysis

This result adequately explains the Gram-determinant singularities but it seems paradoxical that the $\vec{k}$ (which we have in fact simply used as auxiliary variables to help in the calculation of D and its derivations) do not have to correspond to vectors of the diagram lying on the mass shell. The paradox is resolved by the observation that the $\vec{k}$

do not correspond to the values of loop momenta at which the pinch takes place in the $(\vec{k}, \alpha)$ representation of the Feynman integral. To see this, and to find out to what in fact they do correspond, it is necessary to reconsider the singularity in the $(\vec{k}, \alpha)$ representation.

We have already said that we expect that the pinch takes place at infinite values of the loop momenta. In order to make infinity accessible to our analysis we make the transformation

$$\vec{k} = \vec{K}/\zeta, \tag{2.10.19}$$

so that infinity in $\vec{k}$ will be given by $\zeta = 0$. In order to have the correct numbers of new variables one of the components of one of the Lorentz vectors in $\vec{K}$ must be taken constant to compensate for the presence of the new variable ζ. We then consider

$$\Psi = \zeta^2 \psi = \vec{K}^T . \mathbf{A} \, \vec{K} - 2\vec{K}^T . \mathbf{B} \, \vec{p}\zeta + (\vec{p}^T . \mathbf{\Gamma p} - \sigma) \, \zeta^2. \tag{2.10.20}$$

The conditions for a pinch-type configuration are†

$$\frac{\partial \Psi}{\partial \vec{K}} = 0, \quad \frac{\partial \Psi}{\partial \zeta} = 0, \quad \frac{\partial \Psi}{\partial \alpha} = 0. \tag{2.10.21}$$

If $\zeta = 0$, these reduce to

$$\left. \begin{aligned} \mathbf{A} \vec{K} &= 0, \\ \vec{K}^T . \mathbf{B} \vec{p} &= 0, \\ \vec{K}^T . \frac{\partial \mathbf{A}}{\partial \alpha_i} \vec{K} &= 0. \end{aligned} \right\} \tag{2.10.22}$$

These equations have solutions for all choices of external vectors $\vec{p}$, which are obtained by writing

$$\vec{K} = \mathbf{K} \vec{\lambda}. \tag{2.10.23}$$

The α's are chosen to satisfy $C = 0$ so that there is a vector $\mathbf{K}$ such that

$$\mathbf{A K} = 0, \tag{2.10.24}$$

and $\vec{\lambda}$ is a Lorentz vector chosen to satisfy

$$\left. \begin{aligned} \vec{\lambda} . (\mathbf{K}^T . \mathbf{B}\vec{p}) &= 0, \\ \vec{\lambda}^2 &= 0. \end{aligned} \right\} \tag{2.10.25}$$

Then (2.10.23) provides a solution of (2.10.22).

---

† Here we have differentiated with respect to all $\vec{K}$, even the constant component. The apparently inadmissible equation corresponding to this constant component is simply equivalent to the condition $\Psi = 0$, which is also used. This is because Ψ is homogeneous in $\vec{K}$ and ζ. (Cf. the treatment of the α's despite the δ-function making $\Sigma\alpha = 1$.)

A pinch which exists for all values of the external invariants cannot in fact give a singularity. Since the integral is not singular at all points the hypercontour is not trapped at these points. It can then only become trapped as one makes analytic continuation if the permanently existing pinch configuration becomes yet more degenerate. Thus the condition for a singularity is the existence of a 'super pinch', in the formation of which the contour can become trapped. This will happen if we can also find solutions of (2.10.21) at the point

$$\vec{\mathbf{K}} + \delta\vec{\mathbf{K}}, \quad \delta\zeta, \quad \alpha + \delta\alpha, \tag{2.10.26}$$

adjacent to the solution given by (2.10.23–25). The conditions for this are

$$\left.\begin{array}{c}
\mathbf{A}\delta\vec{\mathbf{K}} = \mathbf{B}\vec{\mathbf{p}}\delta\zeta - \delta\mathbf{A}\mathbf{K}\vec{\lambda}, \\[2mm]
-\delta\vec{\mathbf{K}}^T.\,\mathbf{B}\vec{\mathbf{p}} + (\vec{\mathbf{p}}^T.\,\mathbf{\Gamma}\vec{\mathbf{p}} - \sigma)\,\delta\zeta - \vec{\lambda}\mathbf{K}^T.\,\delta\mathbf{B}\vec{\mathbf{p}} = 0, \\[2mm]
\vec{\lambda}\mathbf{K}^T.\,\dfrac{\partial \mathbf{A}}{\partial \alpha_i}\,\delta\vec{\mathbf{K}} - \vec{\lambda}\mathbf{K}^T.\,\dfrac{\partial \mathbf{B}}{\partial \alpha_i}\,\vec{\mathbf{p}}\delta\zeta = 0.
\end{array}\right\} \tag{2.10.27}$$

If we make the identification

$$\vec{\mathbf{k}} = \delta\vec{\mathbf{K}}/\delta\zeta, \tag{2.10.28}$$

the content of (2.10.27) can be made identical with that of our α-space discussion. The first equation of (2.10.27) is identified with (2.10.12). The imposition of the conditions (2.10.13) on $\vec{\lambda}$ reduces the second equation of (2.10.27) to the condition (2.10.18), while the condition that $\vec{k}$ are linear combinations of the $\vec{p}$, together with (2.10.13), satisfies the third equation of (2.10.27). We are thus able to understand why the $\vec{k}$ do not have to correspond to vectors of the diagram on the mass shell. They simply represent the *directions* from which the super pinch at infinity is formed.

We see that the zero-vector solutions of the Gram-determinant conditions are vital in the discussion of second-type singularities. If $E = 5$ these solutions do not exist. There are four independent momenta and the Gram-determinant condition simply makes them lie in a space of three dimensions. Since it requires two dimensions for a zero-length vector it is not possible in four-dimensional space to find a zero-length vector outside this three-dimensional subspace. Thus the $E = 5$ singularities are absent.

The single-loop diagrams require special discussions since C is just the sum of the α's, which is prevented from vanishing by the overall

δ-function. It can be shown (Fairlie *et al.* 1962*a*) that nevertheless there is a second-type singularity for a loop diagram for $E \leqslant 3$ corresponding to the α's tending to infinity.

Mixed second-type singularities

The second-type singularities discussed so far all correspond to the Gram-determinant curve however complicated the Feynman diagram may be. In the integration space they correspond to super pinches at infinity in all the loop momenta and are called *pure second-type* singularities. In a diagram with several loops, however, there may be super pinches only for some of the loop momenta while the others have ordinary pinches at finite points. These singularities are called *mixed second-type* singularities and their equations will depend upon the internal masses of the lines round the loops with finite pinches.

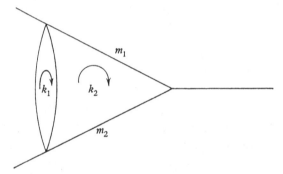

Fig. 2.10.2. An example of a Feynman diagram that gives a mixed second-type singularity.

Their existence can most readily be seen by considering a Feynman integral as being built up by successive integrals over single loops. For example, in the diagram of Fig. 2.10.2, we can perform the k_1-integration over the two-particle loop, to obtain a function which will depend on k_2 and have a second-type singularity. In the k_2-integration this second-type singularity will give a singularity of the complete integral by making a pinch with the poles at m_1^2 and m_2^2 in the remaining lines of the k_2 loop. This is a mixed second-type singularity whose equation can be obtained from the dual diagram of a triangle graph whose internal masses are taken to be m_1, m_2 and 0. It has been investigated in detail by Drummond (1963); see also Fowler (1962).

Little is known about the Riemann-sheet properties of second-type singularities and this remains an important subject for investigation.

CHAPTER 3

ASYMPTOTIC BEHAVIOUR

3.1 Complex angular momentum†

Regge poles

It is a well-known fact that the bound states of the Schrödinger equation for a spherically symmetric potential fall into families characterised by increasing angular momentum and decreasing binding energy. Such a family appears as a sequence of poles occurring in the successive partial wave amplitudes $a_l(s)$, $l = 0, 1, 2, \ldots$, at increasing values of s. The theory of complex angular momentum introduced into potential theory by Regge (1959) pictures this sequence as due to the presence of a single pole whose position varies continuously with l and which is relevant to the physics of bound states only when l takes non-negative integral values. This idea is given meaning by the construction of an interpolating amplitude $a(l, s)$, defined for non-integral, and indeed complex, values of l, which coincides with the physical amplitudes $a_l(s)$ when $l = 0, 1, 2, \ldots$. This function $a(l, s)$ is an analytic function of its arguments except for certain singularities. Among these singularities will be a pole corresponding to each of the bound state families. The location or *trajectory* of such a pole is given by an equation of the form

$$l = \alpha(s). \qquad (3.1.1)$$

Such a singularity is called a *Regge pole*. The bound-state energies will correspond to values of s in (3.1.1) which make l take the values $0, 1, 2, \ldots$.

Watson–Sommerfeld transform

If one such interpolating function $a(l, s)$ can be found then there are many, for we could always add a term

$$f(s) \sin \pi l, \qquad (3.1.2)$$

without altering the properties at physical values of l. What makes the idea of complex angular momentum more than an empty, if elegant,

† This section and the one following summarise results and ideas which will be useful in the sequel but which are not treated exhaustively here. For a fuller account of these topics see Squires (1963).

notion is the fact that there is a particular definition of $a(l,s)$ which enables the interpolating function to be used to derive further results of physical importance.

This arises from the possibility of finding a representation of $A(s,z)$, the scattering amplitude regarded as a function of invariant energy s and cosine of the scattering angle z, which is valid outside the ellipse of convergence of the partial wave series. The technical trick which makes this feasible is called the *Watson–Sommerfeld transform*. Let us suppose that $a(l,s)$ is analytic in $\mathrm{Re}\,l \geqslant L$ for some given value of s. If N is the first integer $> L$ then we may rewrite the partial wave series

$$
\sum_{l=0}^{\infty} (2l+1)\, a_l(s)\, P_l(z)
$$
$$
= \sum_{l=0}^{N-1} (2l+1)\, a_l(s)\, P_l(z) - \frac{1}{2i} \int_C \frac{(2l+1)\, a(l,s)\, P_l(-z)}{\sin \pi l}\, dl, \quad (3.1.3)
$$

where C is the contour shown in Fig. 3.1.1 encircling the integers greater than L. The residues of the integral at the zeros of $\sin \pi l$ just

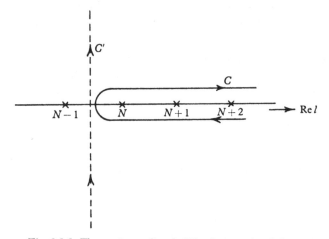

Fig. 3.1.1. The contours C and C' in the complex l-plane.

reproduce the missing terms of the partial wave series. We use the function $P_l(-z) = (-)^l P_l(z)$ to take account of the alternating sign of $(\pi \cos \pi l)^{-1}$, the residue of $(\sin \pi l)^{-1}$ at its poles.

The contour is now opened out to become C', a contour running parallel to the imaginary axis. If $a(l,s)$ is bounded by a power of l it may be shown that the neglect of the contribution from the semicircular arc at infinity is justified for $-1 < z < 1$. Now the integral along C' can be shown to be convergent for all z except possibly $z \geqslant 1$. Thus

$$\sum_{l=0}^{N-1} (2l+1)\, a_l(s)\, P_l(z) - \frac{1}{2i} \int_{C'} \frac{(2l+1)\, a(l,s)\, P_l(-z)}{\sin \pi l}\, dl \qquad (3.1.4)$$

is a representation of $A(s,z)$ valid in the whole z-plane cut along $z \geqslant 1$.

If $a(l,s)$ has suitable analytic properties it will be possible to displace C' further to the left in the complex l-plane. For example, suppose $a(l,s)$ is meromorphic in $L_1 \leqslant \mathrm{Re}\, l < L$ so that its only singularities in this region are Regge poles. Then, as the contour C' is moved over to a new position C_1' running through L_1, additional contributions will be obtained from the residues of the Regge poles that are crossed over, together with residues from the poles of $(\sin \pi l)^{-1}$. These latter terms will just cancel some of the terms in the truncated partial wave series in the right-hand side of (3.1.4). The resulting expression is

$$\sum_{l=0}^{N_1-1} (2l+1)\, a_l(s)\, P_l(z) - \frac{1}{2i} \int_{C_1'} dl\, \frac{(2l+1)\, a(l,s)\, P_l(-z)}{\sin \pi l}$$
$$- \sum_i \frac{(2\alpha_i(s)+1)\, \beta_i(s)\, P_{\alpha_i(s)}(-z)}{\sin \pi \alpha_i(s)}, \qquad (3.1.5)$$

where N_1 is the first integer greater than L_1 and the sum in the third term is taken over Regge poles of trajectory $\alpha_i(s)$ and residue $\pi^{-1}\beta_i(s)$ which have been crossed in the displacement of the contour.

An important deduction can be made from (3.1.5). The Legendre functions have the asymptotic behaviour†

$$P_\alpha(-z) \sim (-z)^\alpha \quad (z \to \infty, \quad \mathrm{Re}\, \alpha \geqslant -\tfrac{1}{2}). \qquad (3.1.6)$$

Therefore the dominant behaviour as $z \to \infty$ will be given by the term in (3.1.5) with the largest value of $\mathrm{Re}\, \alpha$. The Regge pole terms will therefore all be more important than either the truncated series or the integral along C_1' in determining the behaviour as $z \to \infty$ and the dominant term in this behaviour will correspond to the right-most Regge pole.

Thus the theory of complex angular momentum succeeds in forging a remarkable connection between the bound states of the theory and

---

† We are assuming that C_1' does not lie to the left of $\mathrm{Re}\, l = -\tfrac{1}{2}$. The theory can, however, be modified to admit displacements into $\mathrm{Re}\, l < -\tfrac{1}{2}$ (Mandelstam, 1962).

the behaviour at large z. As we shall see in the next section this relationship is particularly important in a relativistic theory as the large z behaviour at fixed s is simply the high energy behaviour in the crossed channel, so that if $s \leqslant 0$ this is a physically observable quantity.

We conclude this section with two remarks. The first is that the utility of complex angular momentum depends upon the feasibility of the Watson–Sommerfeld transform, which in turn requires a bound on the behaviour of $a(l, s)$ as $l \to \infty$ in the right half l-plane. If this condition is satisfied the definition of $a(l, s)$ is in fact unique and no arbitrary terms like (3.1.2) can be added without spoiling the behaviour. This is guaranteed by *Carlson's theorem*. This states that if $a(l, s)$ is analytic in $\mathrm{Re}\, l \geqslant L$ and we require

$$a(l, s) = O(e^{\lambda |l|}) \quad (\lambda < \pi,\ |l| \to \infty,\ \mathrm{Re}\, l \geqslant L), \qquad (3.1.7)$$

then $a(l, s)$ is uniquely defined by its values at the integral values of l.

The second remark is that if singularities other than Regge poles are encountered in displacing the contour then of course they also provide contributions to the high z behaviour. For example a cut c trailed by a singularity at l_0 and having discontinuity $\delta(l, s)$ gives an additional term in (3.1.5)

$$-\frac{1}{2i} \int_{(c)}^{l_0} dl \, \frac{(2l+1)\, \delta(l, s)\, P_l(-z)}{\sin \pi l}. \qquad (3.1.8)$$

3.2 Relativistic theories

The Pomeranchuk pole

In the scattering of particles of equal mass, m, the Mandelstam invariants (s, t, u) are given in terms of k, the centre of mass momentum and z, the cosine of the scattering angle, by the equations

$$s = 4(m^2 + k^2), \quad t = -2k^2(1-z), \quad u = -2k^2(1+z). \qquad (3.2.1)$$

Thus the limit $|z| \to \infty$ at fixed s corresponds to $|t| \to \infty$ at fixed s. However, because of the crossing property, we interpret $t \to \infty$ as the high-energy limit in the crossed channel provided that s is fixed at an appropriate physical value < 0. This, of course, requires analytic continuation in s from the values given by (3.2.1) for real k.

Thus in a relativistic crossing-symmetric theory a Regge pole in the s channel has dual physical significance. It gives the bound states in that channel and it also contributes to the high-energy behaviour in the crossed channel.

The most striking thing about observed high-energy behaviour is the high degree of constancy of total cross-sections for all scattering processes above a few BeV. The total cross-section is related by the optical theorem (1.4.15) to the imaginary part of the forward elastic scattering amplitude. The forward direction in the crossed channel is given by $s = 0$ and the proportionality factor appearing in the optical theorem is a numerical factor times t^{-1}. Thus the constancy of high-energy cross-sections will be explained if all scattering amplitudes are asymptotic like t at $s = 0$. This behaviour represents the maximum permitted by Froissart's result (1.4.13), disregarding a possible logarithmic factor. It would be produced if there were a Regge pole P, called the 'Pomeranchukon', which passed through $l = 1$ when $s = 0$.† Because all high-energy total cross-sections are effectively constant it is necessary that any pair of particles should be able to exchange P. The intrinsic quantum numbers of P are therefore the same as the vacuum (parity $+$, $I = 0$, $S = 0$, etc.).

The behaviour of the amplitude for values of s just below zero determines the form of scattering near the forward direction, that is to say in the high-energy limit it gives the shape of the diffraction peak. If the Pomeranchukon is assumed to be the dominant singularity this shape turns out to be an exponential which shrinks logarithmically with increasing energy. However this simple prediction is neither in accord with experiment nor with more detailed theoretical investigations (see § 3.8). The important problem of determining the form of the diffraction peak remains a complicated and unsolved question.

The Froissart–Gribov continuation

However complicated these questions may be, the way to their answer is clearly in the complex angular momentum plane. In attempting to extend the ideas discussed in § 3.1 to a relativistic theory an immediate difficulty is encountered. In potential theory we are able to use the Schrödinger equation to construct an interpolating function $a(l, s)$ which can be used in a Watson–Sommerfeld transform. No such equation is to hand in a relativistic theory so another method of definition must be found. The answer was provided by the work of Froissart (1961a) and Gribov (1962).

---

† This does not give a spin 1 particle of zero mass if P has positive signature; see below.

We consider for simplicity the scattering of spinless particles of equal mass. At fixed s a dispersion relation is assumed for the scattering amplitude $A(s,t)$ with the integrals taken over the t and u normal threshold cuts:

$$A(s,t) = \frac{1}{\pi}\int_{4m^2}^{\infty}\frac{A_t(s,t')\,dt'}{t'-t} + \frac{1}{\pi}\int_{4m^2}^{\infty}\frac{A_u(s,u')\,du'}{u'-u}. \qquad (3.2.2)$$

We have assumed that no subtractions are necessary. What happens when this is not true will be stated later.

The partial wave projection of (3.2.2) is

$$a_l(s) = \frac{1}{2\pi}\int_{-1}^{1}dz\left[\int_{z_0}^{\infty}dz'\frac{P_l(z)}{z'-z}A_t(s,2k^2(z'-1))\right.$$

$$\left. +\int_{-z_0}^{-\infty}dz'\frac{P_l(z)}{z'-z}A_u(s,2k^2(-z'-1))\right], \qquad (3.2.3)$$

where
$$z_0 = 1 + \frac{2m^2}{k^2}. \qquad (3.2.4)$$

For integral l Neumann's formula,[†]

$$\frac{1}{2}\int_{-1}^{1}dz\frac{P_l(z)}{z'-z} = Q_l(z'), \qquad (3.2.5)$$

makes it possible to rewrite (3.2.3) in the form

$$a_l(s) = \frac{1}{\pi}\int_{z_0}^{\infty}dz'A_t(s,2k^2(z'-1))\,Q_l(z')$$

$$-\frac{1}{\pi}\int_{z_0}^{\infty}dz'A_u(s,2k^2(z'-1))\,Q_l(-z'). \qquad (3.2.6)$$

As $l\to\infty$
$$Q_l(z) \sim l^{-\frac{1}{2}}\exp[l\log\{z-(z^2-1)^{\frac{1}{2}}\}]. \qquad (3.2.7)$$

Therefore with $k^2 > 0$ so that $z_0 > 1$, the first term in (3.2.6) gives sufficient convergence in $\mathrm{Re}\,l > 0$ to permit a Watson–Sommerfeld transform. However, the second term does not have this property owing to the presence of an additional $e^{i\pi l}$ factor which is unbounded as $l\to\infty$ in the lower half plane.

This defect is remedied by defining two functions

$$a^{\pm}(l,s) = \frac{1}{\pi}\int_{z_0}^{\infty}dz'Q_l(z')[A_t(s,2k^2(z'-1)) \pm A_u(s,2k^2(z'-1))]. \qquad (3.2.8)$$

---

† For properties of Legendre functions see Erdelyi et al. (1953).

Both functions have an asymptotic behaviour which permits a Watson–Sommerfeld transform. Because of the identity

$$Q_l(z) = (-)^{l+1} Q_l(-z), \quad l \text{ integral}, \tag{3.2.9}$$

$a^+(l, s)$ coincides with $a_l(s)$ for even l and $a^-(l, s)$ coincides with $a_l(s)$ for odd l.

The need to use distinct continuations for the even and odd angular-momentum states is not surprising. The invariants t and u are obtained from each other by the interchange of the two particles in the final state of the s-channel. Thus if the exchange of particles corresponding to poles in t generates a potential V the exchange of particles corresponding to the same poles in u will generate the corresponding exchange potential V_{ex}. In even angular-momentum states interchanging the two particles produces no change of sign, so the two potentials combine to give an effective potential $V + V_{\text{ex}}$. However, in the odd angular momentum states the interchange produces a minus sign and the effective potential in $V - V_{\text{ex}}$. It is natural, then, that these two different effective potentials will be associated with different families of Regge poles. The label $(\pm)$ is called the *signature*. Even-signature Regge poles only correspond to particles of angular momentum $0, 2, \ldots$, and odd-signature Regge poles only correspond to particles of angular momentum $1, 3, \ldots$.

Under the assumptions that lead to no subtractions in (3.2.2) the integrals of (3.2.8) are convergent for $\operatorname{Re} l > 0$ and so (3.2.8) defines functions $a^\pm(l, s)$ analytic in this region. Had it been necessary to make n subtractions (see Squires, 1963) (3.2.8) would have again been obtained as the appropriate continuation, but $a^\pm(l, s)$ would then only be known to be analytic in $\operatorname{Re} l > n$.

The Gribov–Pomeranchuk phenomena

Two interesting deductions can be made from the Froissart–Gribov definition (3.2.8). The first is called the Gribov–Pomeranchuk (GP) phenomenon (Gribov & Pomeranchuk, 1962a). We shall outline the essential steps in the argument.

The amplitudes $a^\pm(l, s)$ have a number of singularities in s which arise from the singularities of Q_l at $z' = \pm 1$ and the singularities of A_t and A_u producing pinches or end-point singularities in the z' integration. One of these singularities is generated by the singularity possessed by both A_t and A_u which corresponds to the boundary of the third Mandelstam spectral function ρ_{tu}. This is a characteristically relativistic phenomenon since the third spectral function is zero in potential theory, where the t and u channels do not correspond to physical processes. The discontinuity associated with this singularity can be calculated and it is found to consist of an integral over a *finite* range of an integrand involving Q_l.

Now suppose that it is possible to continue $a^{\pm}(l, s)$ to the neighbourhood of $l = -1$. If singularities in l exist to the right of $l = -1$ we shall encounter convergence problems in (3.2.8), so that even though the analytic continuation may be possible we cannot expect that the representation (3.2.8) will be valid for all the continuation. However, the expression for the particular discontinuity we are discussing must remain valid, for there are no convergence problems in the finite-range integral. Therefore this discontinuity, and hence the whole function, can be expected to have a pole at $l = -1$ due to the pole there of the Q_l function appearing in the integrand. In fact more careful investigation show that this pole is cancelled in $a^{+}(l, s)$ but that it is inescapably present in $a^{-}(l, s)$ for all values of s.

An immediate paradox is encountered if this result is applied to values of s chosen to lie in the region of elastic unitarity. Elastic unitarity for physical l may be written in the form (see (1.3.38))

$$a^{\pm}(l, s) - a^{\pm*}(l^*, s) = \frac{i}{8\pi} \sqrt{\frac{s - 4m^2}{s}} \, a^{\pm}(l, s) \, a^{\pm*}(l^*, s). \quad (3.2.10)$$

We have inserted the stars, denoting complex conjugation, in such a way as to make (3.2.10) an analytic function of l. Carlson's theorem will then imply that the equation can be uniquely continued away from integral l. However, if there is a pole at $l = -1$ the left-hand side of (3.2.10) would have a simple pole there, while the right-hand side would have a double pole. This paradox can be resolved in two ways. In the first way there is a cut in l intervening to screen off $l = -1$ from the analytic continuation of (3.2.10) from the non-negative integers. If this cut is crossed a corresponding extra discontinuity must be added so that (3.2.10) must be modified. Alternatively, instead of a pole there is an essential singularity at $l = -1$. We shall return to the discussion of this question in § 3.8.

The second deduction which can be made from (3.2.8) we shall simply state. Gribov & Pomeranchuk (1962 b) deduced from a consideration of the threshold behaviour of the Froissart–Gribov amplitude that there must be an accumulation of an infinite number of

Regge poles along $\mathrm{Re}\, l = -\tfrac{1}{2}$ at the two-particle threshold $s = 4m^2$. We shall encounter this result from a somewhat different point of view in § 3.6.

3.3 High-energy behaviour in perturbation theory

Bound states and the Bethe–Salpeter equation

The region of known analyticity of the Froissart–Gribov (FG) continuation is determined by the subtraction constants. If n subtractions are needed we can only assert that $a^{\pm}(l, s)$ is analytic in $\mathrm{Re}\, l > n$. (The GP phenomenon can then only be deduced on the assumption that continuation to the neighbourhood of $l = -1$ is possible). The singularities in l of the FG continuation to the right of $\mathrm{Re}\, l = -1$ arise from divergences of the z' integration at infinity. Their nature could be determined from a knowledge of the high z' behaviour of A_t and A_u, but it was precisely in the attempt to determine such behaviour that we had recourse to the notion of complex angular momentum! Thus, while the FG continuation strongly suggests that the complex angular momentum plane is the right place to be, it does not help us to determine what we shall find there.

In such a situation the study of appropriate models is a fruitful way of proceeding. The only model which formally reproduces all the properties of an analytic, crossing-symmetric, relativistic theory is that provided by perturbation theory. Of course, the study of perturbation theory is a heuristic method since it could only be made rigorous if the series were uniformly convergent, which is certainly not expected to be the case in strong interaction physics. Gell-Mann & Goldberger (1962) have spoken of the 'laboratory' of Feynman integrals in which we can test the consistency of ideas about the structure of relativistic theories and this has proved a valuable role for perturbation theory. As we shall see, we are encouraged to accept the utility of the method for the study of questions relating to complex angular momentum by the natural way in which Regge poles are found to occur.

It is most convenient to look for complex angular momentum properties in terms of their consequences for high-energy behaviour. This leads then to the study of the high-energy behaviour of Feynman integrals. We shall find it possible to give rigorous procedures for determining the high-energy behaviour of any specific Feynman integral. However, we expect that the properties which are of interest to us will be associated not with single integrals but with infinite sets of Feynman integrals. The reason for this is that Regge poles are connected with the bound states of the theory. In quantum field theory bound states are described by the Bethe–Salpeter equation (Bethe & Salpeter, 1951). Symbolically, this is an integral equation for the amplitude f, describing the bound state, of the form†

$$\underline{}(f)\underline{} = \underline{}(V)\underline{} + \underline{}(V)\underline{}(f)\underline{}, \tag{3.3.1}$$

where V is some set of diagrams describing the 'potential' causing binding. V must be irreducible in the sense that none of its diagrams can be decomposed into two parts joined only by two lines. A formal series solution of (3.3.1) gives

$$\underline{}(f)\underline{} = \underline{}(V)\underline{} + \underline{}(V)\underline{}(V)\underline{} + \dots. \tag{3.3.2}$$

This is physically plausible. In order to maintain the binding of the particles in the bound state V must keep on acting and so f must contain terms corresponding to any number of iterations of V. If V is the 'potential' generated by the exchange of a single particle then (3.3.2) corresponds to the complete set of ladder diagrams illustrated in Fig. 3.3.1. It is from sets of diagrams like these that we expect to

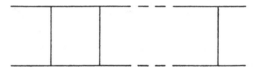

Fig. 3.3.1. A ladder diagram.

extract Regge pole behaviour. Our method will consist in determining the leading asymptotic behaviour for each diagram in the set and then performing a sum. There is no mathematical necessity for the leading asymptotic behaviour of an infinite sum to be the sum of the leading behaviour of its terms, but the answers we shall obtain are heuristically compelling.

Asymptotic behaviour of integrals

First, however, we must consider the asymptotic behaviour of a single Feynman integral (Polkinghorne, 1963*b*; Federbush & Grisaru, 1963). The integral will have the form (see §1.5)

$$\int \Pi \, d\alpha_i \frac{\delta(\Sigma\alpha - 1)\, N(\alpha)}{[g(\alpha)\, t + d(s, \alpha)]^n}, \tag{3.3.3}$$

† In contrast with the bubble equations of chapter 4 the internal lines here represent propagators.

where t is the variable which is to tend to infinity, s is fixed, and the α_i are the Feynman parameters. For the moment we restrict ourselves to similar particles interacting through ϕ^3 interaction, so that the numerator function N is independent of s and t, being just a power of $C(\alpha)$ as in (1.5.22). The extremely interesting features which arise in theories of particles with spin are discussed in § 3.9.

At first sight one might expect from the linear dependence of the denominator function upon t that (3.3.3) would have asymptotic behaviour t^{-n} as $t \to \infty$. This would indeed be so if $g(\alpha)$ did not vanish anywhere in the region of integration. However, as we shall see, an enhanced asymptotic behaviour can be obtained from those parts of the hypercontour of α-integration which are in the neighbourhood of points giving $g = 0$. In order for these points to be really significant it is obviously necessary that the vanishing of g is inescapable and cannot be avoided by the freedom provided by Cauchy's theorem to distort the hypercontour. This means that they must either correspond to g vanishing at the edge of the hypercontour where a number of α's are zero, or they must correspond to g vanishing in a region which is trapped by a pinch as $t \to \infty$. The two different contributions to asymptotic behaviour obtained in these ways are called *end-point contributions* and *pinch contributions*, respectively. In their calculation it is a tremendous simplification that it is only the part of the hyper-contour in the neighbourhood of $g = 0$ which need be considered explicitly.

As a very simple example we consider the integral associated with the square diagram of Fig. 3.3.2, where the α's and β's are the Feynman parameters associated with each line. The integral is

$$g^2 \left(\frac{-g^2}{16\pi^2} \right) \int_0^1 \frac{d\alpha_1 d\alpha_2 d\beta_1 d\beta_2 \, \delta(\alpha_1 + \alpha_2 + \beta_1 + \beta_2 - 1)}{[\alpha_1 \alpha_2 t + d(\alpha, \beta, s)]^2}, \tag{3.3.4}$$

where we exhibit explicitly the dependence of the denominator upon the asymptotic variable t. The coefficient of t vanishes if $\alpha_1 = 0$ or $\alpha_2 = 0$, both of which correspond to edges of the integration region. Since it is only the integration region in the neighbourhood of these edges which will contribute to the dominant term in the high t behaviour we can calculate the leading behaviour by considering instead of (3.3.4) the integral

$$g^2 \left(\frac{-g^2}{16\pi^2} \right) \int_0^\varepsilon d\alpha_1 d\alpha_2 \int_0^1 d\beta_1 d\beta_2 \frac{\delta(\beta_1 + \beta_2 - 1)}{[\alpha_1 \alpha_2 t + d'(\beta, s)]^2}, \tag{3.3.5}$$

where $d'(\beta, s) = d(0, \beta, s).$ (3.3.6)

Equation (3.3.5) has the same leading behaviour as (3.3.4) which will be found to be independent of the value of ϵ in (3.3.5). We now perform the α_1 and α_2 integrations explicitly:

$$\int_0^\epsilon d\alpha_1 d\alpha_2 \frac{1}{[\alpha_1\alpha_2 t + d']^2} = \int_0^\epsilon d\alpha_2 \frac{\epsilon}{d'[\epsilon\alpha_2 t + d']}$$

$$= \frac{1}{d't}\log\frac{\epsilon^2 t + d'}{d'}$$

$$\sim \frac{1}{d'}t^{-1}\log t \quad \text{as} \quad t \to \infty. \qquad (3.3.7)$$

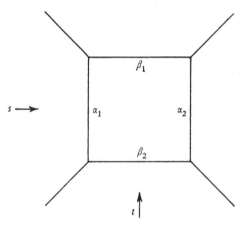

Fig. 3.3.2. The square diagram.

Thus the leading behaviour of (3.3.4) is

$$g^2 K(s)\, t^{-1}\log t, \qquad (3.3.8)$$

where $$K(s) = \frac{-g^2}{16\pi^2}\int_0^1 \frac{d\beta_1 d\beta_2\, \delta(\beta_1 + \beta_2 - 1)}{d'(\beta, s)}. \qquad (3.3.9)$$

The effect of putting an α equal to zero is simply to contract the corresponding line. Therefore the function d' in (3.3.9) is just the Feynman denominator function associated with the contracted diagram, Fig. 3.3.3. The fact that it appears with exponent one shows that the loop momentum associated with Fig. 3.3.3 is to be taken as being two-dimensional rather than four-dimensional.†

† For a four-dimensional loop momentum the diagram is, of course, divergent!

Ladder diagrams

This procedure is easily extended to the n-runged ladder of Fig. 3.3.4. The Feynman integral is of the form

$$g^2 \left(\frac{-g^2}{16\pi^2}\right)^{n-1} (n-1)! \int \frac{\Pi d\alpha d\beta\, \delta(\Sigma\alpha + \Sigma\beta - 1)\, [C(\alpha,\beta)]^{n-2}}{[\alpha_1 \dots \alpha_n t + \delta(\alpha,\beta,s)]^n}. \qquad (3.3.10)$$

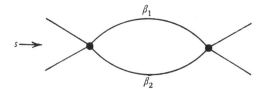

Fig. 3.3.3. The contracted diagram associated with Fig. 3.3.2.

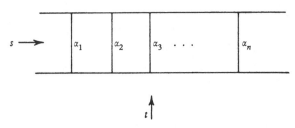

Fig. 3.3.4. The ladder with n rungs.

The α's are the parameters of the rungs of the ladder and the β's are the parameters of the sides; C is the discriminant function of the diagram, as in (1.5.30). The coefficient of t is zero if

$$\alpha_i = 0 \quad (i = 1, \dots, n),$$

corresponding to end-points. Once again, in order to determine the leading behaviour we make the approximation of putting all α's zero except in the vital term involving t, and so consider the integral

$$g^2 \left(\frac{-g^2}{16\pi^2}\right)^{n-1} (n-1)! \int_0^\epsilon d\alpha_1 \dots d\alpha_n \int_0^1 \Pi d\beta\, \frac{\delta(\Sigma\beta - 1)\, [c(\beta)]^{n-2}}{[\alpha_1 \dots \alpha_n t + \delta'(\beta,s)]^n}, \qquad (3.3.11)$$

where $\qquad c(\beta) = C(0,\beta), \quad \delta'(\beta,s) = \delta(0,\beta,s). \qquad (3.3.12)$

In (3.3.11) the α-integrations can again be done explicitly and yield for the leading asymptotic behaviour

$$g^2 t^{-1} \frac{(\log t)^{n-1}}{(n-1)!} \left[\left(\frac{-g^2}{16\pi^2}\right)^{n-1} (n-2)! \int_0^1 \Pi d\beta\, \frac{\delta(\Sigma\beta - 1)\, [c(\beta)]^{n-2}}{[\delta'(\beta,s)]^{n-1}} \right]. \qquad (3.3.13)$$

The functions c and δ' are the correct Feynman numerator and denominator functions for the contracted diagram, Fig. 3.3.5, and the fact that their powers in (3.3.13) differ by one shows that two-dimensional loop momenta must be used in the contracted diagram. Since the diagram is just the product of $(n-1)$ closed loops of the same structure as Fig. 3.3.3 the integral in square brackets must just reduce to

$$[K(s)]^{n-1}. \qquad (3.3.14)$$

Fig. 3.3.5. The contracted diagram associated with Fig. 3.3.4.

A direct analytic proof of this result exhibits some features of interest and is given as an appendix to this section.

The series obtained by summing the leading behaviours of the ladder diagrams is

$$g^2 t^{-1} \sum_{n=1}^{\infty} [K(s) \log t]^{n-1}/(n-1)! = g^2 t^{-1} \exp [K(s) \log t]$$
$$= g^2 t^{\alpha(s)}, \qquad (3.3.15)$$

with
$$\alpha(s) = -1 + K(s). \qquad (3.3.16)$$

The result (3.3.15) is just the behaviour associated with a Regge pole of trajectory (3.3.16). It was first obtained by Lee & Sawyer (1962) by direct consideration of the Bethe–Salpeter equation.

In the next section we shall generalise the method used here for obtaining the leading end-point contribution associated with a Feynman diagram. In §3.5 we shall show how these results lead to Regge pole behaviour. Then in §3.6 we shall give a different method of analysis based on the use of Mellin transforms. This has the great merit that it can be used to derive the lower order terms $(t^{-1} (\log t)^{n-2}$, etc., for the n-runged ladder) as well as the leading term $(t^{-1} (\log t)^{n-1})$. All these terms can be summed for the set of ladder diagrams and shown to correspond to Regge poles. This is a welcome confirmation of the heuristic value of the leading approximation.

Appendix to §3.3

We consider the integral

$$(n-2)! \int_0^1 \Pi d\beta \frac{\delta(\Sigma\beta-1)\,[c(\beta)]^{n-2}}{[\delta'(\beta,s)]^{n-1}} \qquad (A\,3.3.1)$$

appearing in (3.3.13), where c and δ' are associated with Fig. 3.3.5. The rules for forming c (§1.5) immediately imply that it factorises and can be written in the form

$$c = \prod_{j=1}^{n-1}(\beta_j'+\beta_j'') \equiv \prod_{j=1}^{n-1} c_j, \qquad (A\,3.3.2)$$

where β_j' and β_j'' are the Feynman parameters of the jth loop. However δ' has a more complicated structure given by

$$\delta' = \sum_j \prod_{k\neq j} c_k d_j(\beta_j',\beta_j'',s), \qquad (A\,3.3.3)$$

where d_j is the denominator function associated with the jth loop considered by itself.

We now make a set of changes of variables to ρ_j and $\bar{\beta}_j'$ defined by

$$\beta_j' = \rho_j\bar{\beta}_j', \quad \beta_j'' = \rho_j(1-\bar{\beta}_j'). \qquad (A\,3.3.4)$$

The Jacobian of this transformation is ρ_j, so that

$$d\beta_j' d\beta_j'' = \rho_j d\rho_j d\bar{\beta}_j'.$$

This is a particular example of what is called a *scale transformation*, a device which will often be used in subsequent sections. It is convenient formally to rewrite (A 3.3.4) in the form

$$\left.\begin{aligned}\beta_j' = \rho_j\bar{\beta}_j', \quad \beta_j'' = \rho_j\bar{\beta}_j'', \\ d\beta_j' d\beta_j'' \to \rho_j d\rho_j d\bar{\beta}_j' d\bar{\beta}_j'' \,\delta(\bar{\beta}_j'+\bar{\beta}_j''-1).\end{aligned}\right\} \qquad (A\,3.3.5)$$

Also we note that the structure of d_j is such that $d_j \to \rho_j^2 d_j$.

The result of these substitutions in (A 3.3.1) produces

$$(n-2)! \int_0^1 \Pi d\bar{\beta} \int_0^1 \Pi d\rho \frac{\delta(\Sigma\rho-1)\prod\limits_{j=1}^{n-1}\delta(\bar{\beta}_j'+\bar{\beta}_j''-1)}{[\sum\limits_j \rho_j d_j(\bar{\beta}_j',\bar{\beta}_j'',s)]^{n-1}}. \qquad (A\,3.3.6)$$

Using the Feynman identity

$$(n-2)! \int_0^1 d\rho_j \frac{\delta(\Sigma\rho-1)}{[\Sigma\rho_j d_j]^{n-1}} = \prod_{j=1}^{n-1} \frac{1}{d_j} \qquad (A\,3.3.7)$$

(A 3.3.6) reduces to

$$\prod_{j=1}^{n-1}\left\{\int_0^1 d\bar{\beta}'_j \bar{d}\beta''_j \frac{\delta(\bar{\beta}'_j + \bar{\beta}''_j - 1)}{d_j(\bar{\beta}'_j, \bar{\beta}''_j, s)}\right\}. \qquad (\text{A}\,3.3.8)$$

3.4   End-point contributions

d-lines

In this section we shall extend the ideas on end-point contributions which were illustrated in § 3.3 by consideration of the ladder diagrams. There is an important class of diagrams, called *planar diagrams*

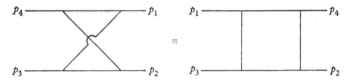

Fig. 3.4.1. A diagram that is non-planar in one limit but not in another.

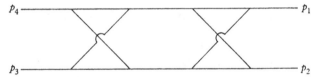

Fig. 3.4.2. A diagram that is non-planar in every limit.

whose physical high-energy behaviour is entirely determined by end-point contributions. The class is defined in the following way. If $p_1, \ldots, p_4$ are the external momenta of a diagram and

$$s = (p_1 + p_2)^2 = (p_3 + p_4)^2 \quad \text{and} \quad t = (p_2 + p_3)^2 = (p_4 + p_1)^2,$$

then for the limit $t \to \infty$ at fixed s the diagram is called planar if it can be drawn in a plane without either internal or external lines crossing, the external lines being attached round the diagram in the order p_1, p_2, p_3, p_4. Otherwise it is called *non-planar*. Notice that the definition depends upon the limit being considered, since this defines the order in which the external lines are attached. For example, the diagram of Fig. 3.4.1 is non-planar for the limit $t \to \infty$ at fixed s, but planar if we consider $t \to \infty$ at fixed u. However diagrams like Fig. 3.4.2 which have all three Mandelstam spectral functions non-zero are non-planar for every limit.

For planar diagrams it follows from the rules of § 1.5 that the coefficient of t, the asymptotic variable, is composed of the *sum* of products of α's. For the positive α's relevant to physical-region high-energy behaviour such an expression can only vanish if a number of α's are zero and so we are restricted to end-point contributions. For non-planar diagrams, however, the coefficient of t may contain products of α's appearing with either sign and such expressions can then vanish in the interior of the region of integration, even for positive α's. This leads to the pinch contributions associated with non-planar diagrams, which we discuss in §§ 3.7 and 3.8.

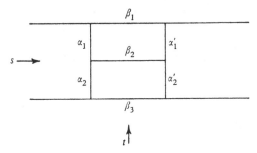

Fig. 3.4.3. A diagram considered in the text.

As our first more complicated example of an end-point contribution we consider the leading behaviour associated with the planar diagram of Fig. 3.4.3. The coefficient of t is

$$g = \alpha_1\alpha_1'(\alpha_2+\alpha_2'+\beta_2+\beta_3) + \alpha_2\alpha_2'(\alpha_1+\alpha_1'+\beta_1+\beta_2)$$
$$+ \alpha_1\beta_2\alpha_2' + \alpha_2\beta_2\alpha_1'. \quad (3.4.1)$$

In order to make g vanish, the least number of parameters it is necessary to put equal to zero is two, and there are two possible pairs for the purpose, α_1, α_2 or α_1', α_2'.

In order to investigate the consequent high-energy behaviour it is convenient to use the scaling techniques introduced by Federbush & Grisaru (1963). Instead of the variables α_1, α_2 we use the new variables ρ and $\bar{\alpha}_1$ defined by $\quad \alpha_1 = \rho\bar{\alpha}_1, \quad \alpha_2 = \rho(1-\bar{\alpha}_1), \quad\quad (3.4.2)$

the Jacobian of the transformation being ρ. In fact it is convenient to introduce formally the further variable $\bar{\alpha}_2 = (1-\bar{\alpha}_1)$ and then write

$$d\alpha_1 d\alpha_2 \to \rho\, d\rho\, d\bar{\alpha}_1 d\bar{\alpha}_2 \delta(\bar{\alpha}_1+\bar{\alpha}_2-1). \quad (3.4.3)$$

The merit of this transformation is that the effect of putting α_1 and α_2 simultaneously zero is now reproduced by setting the single scaling parameter ρ equal to zero. A similar scaling parameter ρ' is introduced for α_1' and α_2'. The coefficient of t may be written

$$g = \rho\rho'\tilde{g}. \tag{3.4.4}$$

In order to extract the leading high-energy behaviour it is sufficient to make the approximation of putting $\rho = \rho' = 0$ everywhere in the integrand except in g, but even here we can put $\rho = \rho' = 0$ in $\tilde{g}$. This produces a new effective coefficient of t of the form

$$\rho\rho'\bar{g} = \rho\rho'[\bar{\alpha}_1\bar{\alpha}_1'(\beta_2+\beta_3)+\bar{\alpha}_2\bar{\alpha}_2'(\beta_1+\beta_2)+\bar{\alpha}_1\bar{\alpha}_2'\beta_2+\bar{\alpha}_2\bar{\alpha}_1'\beta_2]. \tag{3.4.5}$$

The construction of $\bar{g}$ from $\tilde{g}$ is called *linearization*.

The high-energy behaviour now follows from the formula

$$\int_0^\epsilon \rho\, d\rho \int_0^\epsilon \rho'\, d\rho' \frac{1}{[\rho\rho'\bar{g}t+d]^3} \sim \tfrac{1}{2}t^{-2}\log t\, \frac{1}{\bar{g}^2 d}, \quad \text{as} \quad t\to\infty, \tag{3.4.6}$$

which is easily proved by direct integration. Applied to the diagram, Fig. 3.4.3, this yields an asymptotic behaviour

$$K'(s)\, t^{-2}\log t, \tag{3.4.7}$$

where

$$K'(s) = \int_0^1 \Pi d\beta\, \delta(\Sigma\beta - 1) \int_0^1 d\bar{\alpha}_1 d\bar{\alpha}_2\, \delta(\bar{\alpha}_1+\bar{\alpha}_2-1) \int_0^1 d\bar{\alpha}_1' d\bar{\alpha}_2'\, \delta(\bar{\alpha}_1'+\bar{\alpha}_2'-1)$$

$$\times \frac{c(\beta)}{[\bar{\alpha}_1\bar{\alpha}_1'(\beta_2+\beta_3)+\bar{\alpha}_2\bar{\alpha}_2'(\beta_1+\beta_2)+\bar{\alpha}_1\bar{\alpha}_2'\beta_2+\bar{\alpha}_2\bar{\alpha}_1'\beta_2]^2\, d(\beta,s)}, \tag{3.4.8}$$

with c and d are the Feynman functions of the contracted diagram, Fig. 3.4.4.

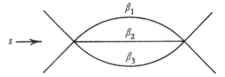

Fig. 3.4.4. The contracted diagram associated with Fig. 3.4.3.

Notice that sets of Feynman parameters that, when zero, make g vanish do not have any effect on the leading behaviour of Fig. 3.4.3 if they contain more than the minimum two parameters. For example, g is zero if $\alpha_1 = \beta_2 = \alpha_2' = 0$. However, the fact that the integral in (3.4.8) is convergent in the neighbourhood of $\bar{\alpha}_1 = \beta_2 = \bar{\alpha}_2' = 0$ shows that scaling their variables would not further enhance the leading behaviour.

The expression (3.4.6) can be generalised to give

$$\int_0^\epsilon \rho_1^{n-1} d\rho_1 \dots \int_0^\epsilon \rho_m^{n-1} d\rho_m \frac{1}{[\rho_1 \dots \rho_m \bar{g} t + d]^p}$$

$$\sim \frac{(n-1)!\,(p-n-1)!}{(p-1)!} \frac{1}{\bar{g}^n d^{p-n}} t^{-n} \frac{(\log t)^{m-1}}{(m-1)!} \quad (t \to \infty,\ p > n). \quad (3.4.9)$$

If $p = n$ the leading asymptotic behaviour of (3.4.9) $\sim t^{-n}(\log t)^m$, and if $p < n$ the leading behaviour $\sim t^{-p}$.

The scaling procedure also generalises. If we wish to set $\alpha_1, \dots, \alpha_n$ simultaneously zero we introduce the variables ρ, $\bar{\alpha}_i$ and make the transformation

$$\left.\begin{aligned} \alpha_i &= \rho \bar{\alpha}_i \quad (i = 1, \dots, n), \\ d\alpha_1 \dots d\alpha_n &\to \rho^{n-1} d\rho\, d\bar{\alpha}_1 \dots d\bar{\alpha}_n\, \delta(\Sigma\bar{\alpha} - 1). \end{aligned}\right\} \quad (3.4.10)$$

These expressions suggest the following approach to determining end-point contributions. We look in the diagram under consideration for connected open arcs of lines which when they are contracted reduce the diagram into two vertex parts in s joined through the single point corresponding to the contracted arc. These arcs correspond to sets of α's which make the coefficient of t vanish when they are set equal to zero. Moreover, we only consider those arcs which are of minimum length, that is they are composed of the least possible number of lines. Such sets of lines are called *d-lines* (Halliday, 1963) or *t-paths* (Tiktopoulos, 1963a). For example, the diagram figure (3.4.3) has two d-lines, each of length 2, corresponding to the sets of parameters α_1, α_2; α_1', α_2'. If one associates a scaling parameter ρ_i with each d-line then (3.4.9) suggests that the leading behaviour will have a power of t (t^{-n}) given by the length of the d-lines, and a power of the logarithm $([\log t]^{m-1})$ one less than their number. This is indeed correct if the residual integral obtained after performing these scalings is convergent. However, this is not always the case and we shall go on to illustrate the circumstances in which more complicated calculations are necessary.†

Singular configurations

Tiktopoulos (1963a) pointed out that in addition to d-lines it is also necessary to consider certain other sets of lines called *singular configurations*. They are distinguished from d-lines by the fact that some

---

† We need not, however, give a complete set of codified rules for leading behaviour, as these are rather complicated to state and in all cases of interest it is easier to apply the ideas directly to the diagrams being considered. Rules were given by Tiktopoulos (1963a) but, as we shall see, they need some amendment.

of the lines composing them form closed loops. Their importance arises
in the following way.

In a theory of spinless particles interacting through a ϕ^3 term in the
Lagrangian, the powers of the Feynman functions $C(\alpha)$ and $D(\alpha; s, t)$
always differ by two, giving an integrand of the form

$$\frac{[C(\alpha)]^{p-2}}{[D(\alpha; s, t)]^p}. \tag{3.4.11}$$

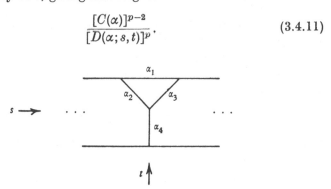

Fig. 3.4.5. A singular configuration.

If the α's round a loop are all made to vanish like ρ (the appropriate
scaling parameter) then C and D both vanish linearly like ρ also. The
resulting net power of ρ^{-2} reduces the effective length of scalings
containing a single closed loop by two. For example, consider
Fig. 3.4.5 considered as an insertion occurring in a more complicated
diagram. Suppose we scale with ρ over the whole set of parameters
$\alpha_1, \ldots, \alpha_4$. The resulting integral has the form

$$\int \rho \, d\rho \, d\bar{\alpha} \ldots \frac{[C(\alpha, \bar{\alpha})]^{p-2}}{[\rho \bar{g} t + d]} \delta(\Sigma \bar{\alpha} - 1), \tag{3.4.12}$$

where the net power of ρ is obtained from a combination of ρ^3 in the
Jacobian with the ρ^{-2} extracted from numerator and denominator.
This scaling will be as effective in producing an enhanced asymptotic
behaviour as are the d-line scalings over the sets α_2, α_4; α_3, α_4. More
generally, if we have a d-line of length n and by adding to it $2m$ further
lines we can form m closed loops then scaling over this augmented set
is as important as scaling over the original d-line, whilst if m closed
loops could be formed by the addition of $2m - 1$ (or less) lines then the
scaling over the augmented set would give an asymptotic behaviour
with one (or more) fewer powers of t than the original d-line scaling.

In fact the singular configurations which are possible in a ϕ^3 theory
can be classified into the three groups of Fig. 3.4.6. Those of class (a)

give an asymptotic behaviour with one less inverse power of t than is given by scaling the two d-lines of the insertion. Those of class (b) (formed by adding a line to class (a)), and those of class (c), give scalings comparable to the d-line scalings and so enhance the power of the logarithm obtained.

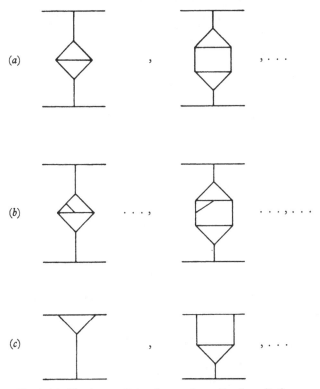

Fig. 3.4.6. The types of singular configuration in a ϕ^3 theory.

Disconnected scalings

It is instructive to consider in greater detail the singular configuration given by the second diagram of class (c). Fig. 3.4.7 labels the Feynman parameters. The factor in the coefficient of t corresponding to this insertion is

$$g = \alpha_1[\beta_1(\alpha_2+\beta_2+\delta_2)+\delta_2\beta_2]+\alpha_2[\beta_2(\alpha_1+\beta_1+\delta_1+\delta_2)+\delta_2\beta_1]$$
$$+\gamma[(\alpha_2+\beta_2+\delta_2)(\alpha_1+\beta_1+\delta_1+\delta_2)-\delta_2^2]. \quad (3.4.13)$$

There are two d-lines of length three and two loops forming singular configurations with them. Therefore we might expect that a complete sequence of scalings would be given by

$$
\left.
\begin{aligned}
&\rho_1\colon \alpha_1,\ \alpha_2,\ \beta_1,\ \beta_2,\ \delta_1,\ \delta_2,\ \gamma;\\
&\rho_2\colon \alpha_1,\ \alpha_2,\ \beta_2,\ \delta_2,\ \gamma;\\
&\rho_3\colon \alpha_1,\ \alpha_2,\ \gamma;\\
&\rho_4\colon \beta_1,\ \beta_2,\ \gamma.
\end{aligned}
\right\}
\tag{3.4.14}
$$

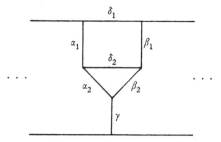

Fig. 3.4.7. A singular configuration having a disconnected scaling.

However, after (3.4.14) has been performed the linearized residual factor $\bar{g}$ is given by

$$
\bar{g} = \bar{\alpha}_1\bar{\delta}_2\bar{\beta}_1 + \bar{\alpha}_2[\bar{\delta}_2\bar{\beta}_1 + \bar{\delta}_1\bar{\beta}_2] + \bar{\gamma}\bar{\delta}_1\bar{\delta}_2,
\tag{3.4.15}
$$

where the parameters are required to satisfy the constraints imposed by the scaling δ-functions:

$$
\delta(\bar{\delta}_1 - 1)\,\delta(\bar{\delta}_2 - 1)\,\delta(\bar{\alpha}_1 + \bar{\alpha}_2 - 1)\,\delta(\bar{\beta}_1 + \bar{\beta}_2 + \bar{\gamma} - 1).
\tag{3.4.16}
$$

Menke (1964) pointed out that in $\bar{g}$ a further scaling is permitted by (3.4.16), which is given by

$$
\rho_5\colon \bar{\alpha}_2,\ \bar{\beta}_1,\ \bar{\gamma}.
\tag{3.4.17}
$$

This scaling corresponds to a *disconnected* set of lines in the original diagram which have become a possible scaling set because of the linearization in forming $\bar{g}$. The possibility of this occurring was first pointed out by Greenman (1965) in the context of fixed angle high-energy limits.

Independent scaling sets

One further point must be made about the evaluation of end-point contributions. It concerns not the type of asymptotic behaviour obtained but the numerical coefficient which multiplies the powers of t

and $\log t$. It may most simply be illustrated by considering the non-planar insertion of Fig. 3.4.8.† The factor in the coefficient of t associated with this insertion is

$$g = \alpha_1 \alpha_2 - \beta_1 \beta_2. \tag{3.4.18}$$

We may begin a sequence of scaling with, say,

$$\rho_1 \colon \alpha_1,\, \beta_1, \tag{3.4.19}$$

with its associated δ-function

$$\delta(\bar{\alpha}_1 + \bar{\beta}_1 - 1). \tag{3.4.20}$$

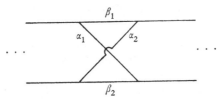

Fig. 3.4.8. A non-planar insertion.

If we take as our second scaling

$$\rho_2 \colon \bar{\alpha}_1,\, \beta_2 \tag{3.4.21}$$

then linearization in (3.4.20) will produce a δ-function requiring

$$\bar{\beta}_1 = 1. \tag{3.4.22}$$

The scaling sequence can then be completed by

$$\rho_3 \colon \bar{\beta}_2,\, \alpha_2. \tag{3.4.23}$$

However we might have taken as our second scaling

$$\rho'_2 \colon \bar{\beta}_1,\, \alpha_2, \tag{3.4.24}$$

which would then have produced from (3.4.20) the condition

$$\bar{\alpha}_1 = 1, \tag{3.4.25}$$

and the sequence would be completed by

$$\rho'_3 \colon \bar{\alpha}_2,\, \beta_2. \tag{3.4.26}$$

It is clear from the equations (3.4.22) and (3.4.25) that these two scaling sequences are independent alternatives, and in calculating the leading asymptotic behaviour both must be taken into account. This

† The effect is also present in the insertion of Fig. 3.4.7, but in a somewhat more complicated form.

fact was first pointed out in the context of production processes (Polkinghorne, 1965 a; see § 3.10). A systematic account has been given by Hamprecht (1965). He has also pointed out that linearization may lead to erroneous calculation of numerical coefficients. A very simple example of this is provided by the integral

$$\int_0^1 \frac{dx\,dy}{(xy+x^2)\,t+1}. \tag{3.4.27}$$

The coefficient of t vanishes if $x = 0$, and if the x^2 term is neglected, according to the linearization idea, the linearised coefficient also vanishes when $y = 0$. This would seem to lead to an asymptotic behaviour $t^{-1}\log t$. In fact explicit integration readily shows that the correct asymptotic behaviour is $\frac{1}{2}t^{-1}\log t$. Care is necessary about the numerical factor only in cases, like (3.4.27), when linearization produces scalings not present in the original coefficient of t.

The ideas of this section have also been applied to high-energy behaviour at fixed angle (when both the energy and the momentum transfer become infinite). Halliday (1964) has discussed the scattering amplitude and Greenman (1965) production processes, both in a ϕ^3 theory.

3.5 Regge poles in perturbation theory

Generalised ladders

In this section we shall show how the application of the idea of the preceding section to simple sets of diagrams leads to end point contributions corresponding to Regge poles. We shall also obtain some properties of the resulting trajectory functions.

As a first illustration we consider the set of diagrams generated by making more complicated insertions between some, but not necessarily all, of the rungs of ladder diagrams (Polkinghorne, 1963 b). An example of this type of diagram is given in Fig. 3.5.1. The new insertions do not provide any further d-lines of length 1 so that the leading end-point high-energy behaviour is still obtained by integrating, in the neighbourhood of the origin, the parameters corresponding to the single-line rungs of the ladder. If there are n such rungs the high-energy behaviour will $\sim t^{-1}(\log t)^{n-1}/(n-1)!$ and the coefficient will be a Feynman integral associated with a diagram formed by contracting the n rungs. Fig. 3.5.2 shows the contracted diagram associated with Fig. 3.5.1. Because these contracted diagrams correspond to a sequence of subdiagrams joined only at common vertices, the corresponding integrals can be decomposed into a product of integrals associated with each subdiagram (cf. the appendix to § 3.3). Thus each possible insertion (labelled by i) will give an integral K_i, associated with a self-energy diagram obtained by joining together both rungs

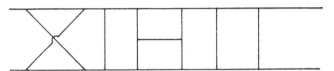

Fig. 3.5.1. A diagram belonging to the class considered.

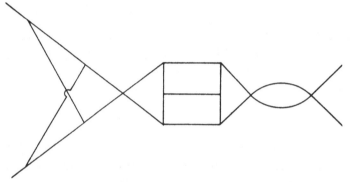

Fig. 3.5.2. The contracted diagram associated with Fig. 3.5.1.

adjacent to the insertion, if it is an internal insertion; and integrals V_i and V_i', associated with vertex diagrams obtained by joining together one or other pair of its external lines, if it appears at one or other end of the ladder. When no insertion is made the corresponding contributions are $K(s)$ (equation (3.3.9)) in the interior, and g at either end. For a given set of insertions, the number of distinct orders in which they can be inserted within the n rungs of the ladder is just a multinomial coefficient. Thus if one sums on all possible insertions in the n-rung ladder the total asymptotic behaviour is given by

$$(g + \Sigma V_i(s))\,(K(s) + \Sigma K_i(s))^{n-1}\,(g + \Sigma V_i'(s))\,t^{-1}(\log t)^{n-1}/(n-1)!, \quad (3.5.1)$$

the sums being taken over all possible insertions i. When (3.5.1) is summed over n it gives a Regge pole behaviour of the form

$$b(s)\,t^{\alpha(s)}, \quad (3.5.2)$$

where

$$\alpha(s) = -1 + K(s) + \Sigma K_i(s), \quad (3.5.3)$$

and

$$b(s) = (g + \Sigma V_i(s))\,(g + \Sigma V_i'(s)). \quad (3.5.4)$$

The terms in the sum in the trajectory (3.5.3) clearly correspond to higher-order corrections in g to the trajectory (3.3.9) which was calculated only to order g^2. However (3.5.3) does not contain all such contributions. Others are associated with the less-than-leading terms. For the simple ladders we shall evaluate these in § 3.6.

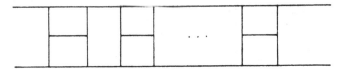

Fig. 3.5.3. The iteration of H-insertions.

Another instructive example is provided by the iteration of the H-insertion of Fig. 3.5.3 (Federbush & Grisaru, 1963). If there are n H's there are $2n$ d-lines of length 2 formed by the sides of the H's. The resulting leading asymptotic behaviour is easily found to be

$$[K'(s)]^n [K(s)]^{n-1} t^{-2} (\log t)^{2n-1}/(2n-1)!, \tag{3.5.5}$$

with $K'(s)$ defined by (3.4.8). When (3.5.5) is summed over n only odd powers of $\log t$ are obtained. This means that the asymptotic behaviour is that associated with a *pair* of Regge poles:

$$\tfrac{1}{2}[K'(s)/K(s)]^{\frac{1}{2}} [t^{\alpha_\pm(s)}], \tag{3.5.6}$$

with
$$\alpha_\pm = -2 \pm [K'(s) K(s)]^{\frac{1}{2}}. \tag{3.5.7}$$

If a general discussion of Regge pole generation is attempted it is in the first instance natural to think of diagrams with sets of d-lines not having a line in common. Then, if these d-lines are scaled over, the contracted diagram will reduce to a set of subdiagrams joined only at common vertices. It will then be necessary to show that the corresponding residual integral can be factorised into a product of integrals, each corresponding to one of the subdiagrams. The factors C and D appearing in the residual integrals can be dealt with by scaling over the parameters of the subdiagrams in the way described in the appendix to § 3.3. The difficult part of the proof is to establish in the general case a corresponding factorisation for the linearized coefficient $\bar{g}$ which remains after scaling. This has been trivial in the examples we have discussed but it is far from obvious in the general case. This factorisation has been established by Halliday (1963) for the case of d-lines and extended to the singular configurations of ϕ^3 theory by Menke (1964). For details of the argument reference should be made to these papers.

Fixed cuts

If end-point behaviour is discussed in a ϕ^4 theory the singular configurations are much more numerous than in ϕ^3 theory and in some cases they lead to asymptotic behaviour different from that associated with Regge poles. Bjorken & Wu (1963) have discussed the series of diagrams illustrated in Fig. 3.5.4. In order to extract the leading behaviour it is necessary to scale over the triangular loops as well as the individual rungs.† The resulting series is found to sum to a

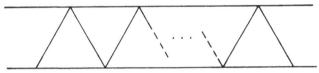

Fig. 3.5.4. The truss-bridge diagrams.

behaviour corresponding in the angular momentum plane to a fixed cut running to the left from $l = -1 + 2\sqrt{G}$, where G is the ϕ^4 coupling constant. Such cuts are also familiar in potential theory where they are associated with the singular nature of the potential (Oehme, 1962; Challifour & Eden, 1963).

Singularities of trajectory functions

We have seen in equation (3.5.3) that when trajectory functions are calculated in perturbation theory the resulting integrals are very similar to those associated with self-energy parts. They will, therefore, have the singularities in s corresponding to normal thresholds. This is to be expected on general grounds since the Froissart–Gribov continuation $a_l^\pm(s)$ has these singularities. However, the trajectory integrals are more complicated than self-energy integrals, owing to the appearance in their integrands of extra factors associated with $\bar{g}$ (cf. equation (3.4.8)). These extra terms will produce additional singularities in s (Polkinghorne, 1963c).

The general theory may be illustrated by the example of the pair of poles generated by the iteration of Fig. 3.5.5. To avoid divergence

---

† Of course the overall δ-function requires one loop to remain unscaled.

difficulties we shall take two-dimensional loop momenta. The trajectory functions of the pair of poles then contains the integral

$$\int_0^1 \frac{\Pi d\gamma \, \delta(\Sigma\gamma - 1) \, [c(\gamma)]^2}{\bar{g}(\gamma) \, [d(s, \gamma)]^2}, \qquad (3.5.8)$$

where

$$\bar{g}(\gamma) = \gamma_1\gamma_2 - \gamma_3\gamma_4, \qquad (3.5.9)$$

and c and d are the Feynman functions associated with the self-energy part obtained by contracting the lines with parameters α_1 and α_2 in Fig. 3.5.5. In addition to the normal-threshold and second-type

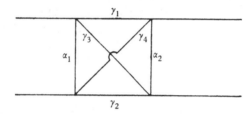

Fig. 3.5.5. A diagram associated with a singularity of a Regge trajectory.

singularities associated with pinches generated by the vanishing of d alone, there is also a singularity generated by $\bar{g}$ and d whose location is determined by the equations

$$d = 0 = \bar{g},$$

$$\frac{\partial d}{\partial \gamma_i} + \lambda \frac{\partial \bar{g}}{\partial \gamma_i} = 0 \quad (i = 1, ..., 4), \qquad (3.5.10)$$

(cf. §2.1). It can be shown that these equations give a value of s identical with a singularity of the Regge amplitude discovered by Islam, Landshoff & Taylor (1963) by considering the asymptotes of the leading Landau curve for Fig. 3.5.5. Notice that in our discussion this singularity is associated with a definite pair of trajectories.

The connection between singularities of trajectories associated with the vanishing of g and asymptotes to Landau curves is quite general (Polkinghorne, 1963c). It can be easily understood in terms of the Froissart–Gribov continuation. Singularities of A_l and A_u occurring at $z = \infty$ give end-point singularities in (3.2.8). Indeed this is how s normal thresholds appear in $a^{\pm}(l, s)$, since A_l and A_u do not have the s normal thresholds themselves.

These new singularities of trajectories will only be present in physical-region behaviour if we can have $g = 0$ for positive α's. They are, therefore, a special feature of non-planar diagrams. They constitute the first of several properties we shall encounter associated with non-planar diagrams which have no analogue in potential theory.

3.6 Mellin transforms and ladder diagrams

General ideas

In this section we introduce the extremely useful Mellin transform technique (Bjorken & Wu, 1963; Trueman & Yao, 1963; Polkinghorne, 1964a) for calculating high-energy behaviour. Its merit, as we shall see, lies in the fact that it provides a method powerful enough to calculate all terms in the asymptotic behaviour. The approximation methods we have used so far are only capable of calculating the leading behaviour.

We shall then go on to use the method to calculate and sum *all* the terms of the form $t^{-1}(\log t)^n$ associated with ladder diagrams. The argument necessarily involves some combinatorial complexities and it may be omitted by the less dedicated reader. However, the result that the final sum corresponds to Regge pole behaviour is important, for it strongly confirms the heuristic value of the methods we use in the perturbation-theory model of high-energy behaviour.

The Mellin transform $F(\beta)$ of a function $f(t)$ is defined by the relation

$$F(\beta) = \int_0^\infty f(t)\, t^{-\beta-1} dt. \tag{3.6.1}$$

It possesses an inversion formula

$$f(t) = \frac{1}{2\pi i} \int_C F(\beta)\, t^\beta d\beta, \tag{3.6.2}$$

where the contour C is parallel to the imaginary β-axis and $F(\beta)$ is analytic along C. Particularly important examples of these relationships are provided by functions of the form

$$f(t) = \frac{t^{\beta_0} (\log t)^{n-1}}{(n-1)!}, \tag{3.6.3}$$

whose Mellin transforms are multiple poles

$$F(\beta) = \frac{1}{(\beta - \beta_0)^n}. \tag{3.6.4}$$

Mellin transforms may be used in a way similar to the use of the Watson–Sommerfeld transformation discussed in §3.1. If $F(\beta)$ is analytic except for multiple poles in some region then we may displace the contour C to the left and obtain a series of contributions from the multiple poles which have been crossed over. The asymptotic behaviour associated with these multiple poles, and also with the integral over the displaced contour, will depend upon the corresponding values of $\mathrm{Re}\,\beta$. The dominant contributions will come from the rightmost singularity in the β-plane.

The utility of the method lies in the fact that simple formulae can easily be found for the Mellin transforms of planar Feynman diagrams. Such diagrams have cuts only in s and t and if we choose the fixed value of s to be below the lowest s cut then the limit $t \to -\infty$ will be a limit taken in a region free from singularity. (We cannot take $t \to +\infty$ since we would then encounter the t normal thresholds.) This is important for the purpose of Mellin transforms since (3.6.1) requires for its validity that the integral should not encounter a singularity.

Our Feynman integral can be written in the form (see (1.5.22))

$$GT(m)\int_0^1 \Pi\,d\alpha\,\frac{\delta(\Sigma\alpha-1)\,[C(\alpha)]^{m-2}}{[D(\alpha,s,t)]^m}, \qquad (3.6.5)$$

where G is a coupling constant factor; $\Gamma(m)$ is the Gamma function, which for integral m equals $(m-1)!$. The value of m is related to the number of lines, p, and the number of independent loops, L, in the diagram by the formula

$$m = p - 2L, \qquad (3.6.6)$$

while C and D are homogenous functions of degrees L and $L+1$ respectively in the α's. We write D in the form

$$D = -g(\alpha)\,\tau - J(\alpha,s)\,C(\alpha), \qquad (3.6.7)$$

where

$$\tau = -t \qquad (3.6.8)$$

is the variable with respect to which we wish to take our Mellin transform. The fact that the limit $\tau \to \infty$ with s below its cuts takes place in a singularity-free region finds its reflection in the fact that, for positive α's, D is negative-definite in this region. C is, of course, positive definite for positive α's. It is then possible to replace (3.6.5) by the expression

$$(-1)^m\,G\int_0^\infty \Pi\,d\tilde{\alpha}\,[C(\tilde{\alpha})]^{-2}\exp\,[D(\tilde{\alpha},s,\tau)/C(\tilde{\alpha})]. \qquad (3.6.9)$$

The equivalence between (3.6.5) and (3.6.9) may be shown by making the scale transformation

$$\tilde{\alpha}_i = \rho \alpha_i \quad (i = 1, 2, ..., p), \tag{3.6.10}$$

and performing the ρ integration with the aid of the formula

$$\Gamma(m) = \int_0^\infty \rho^{-m-1} e^{-\rho} d\rho. \tag{3.6.11}$$

The negative definite property of D makes the ρ integral well-defined. The Mellin transform of (3.6.9) is given by

$$L(\beta, s) = (-1)^m G \int_0^\infty d\tau \int_0^\infty \Pi d\tilde{\alpha}\, \tau^{-\beta-1} \exp\left[-g(\tilde{\alpha})\,\tau/C(\tilde{\alpha})\right]$$
$$\times \exp\left[-J(\tilde{\alpha}, s)\right]. \tag{3.6.12}$$

Interchange of the order of integration and the use again of (3.6.11) enables the t-integration to be performed to give

$$L(\beta, s) = \Gamma(-\beta)(-1)^m G \int_0^\infty \Pi\, d\tilde{\alpha}\, [g(\tilde{\alpha})]^\beta\, [C(\tilde{\alpha})]^{-2-\beta} e^{-J}. \tag{3.6.13}$$

The asymptotic behaviour will be given by a knowledge of the singularities of $L(\beta, s)$ in the left half β-plane. The exponential guarantees convergence at infinity so that the singularities of (3.6.13) will be given by divergences of the integral at the lower limits of integration. These will correspond to the region where some α's vanish: that is, they correspond to the expected end-point behaviour.

Ladder diagrams

We shall illustrate how they can be analysed by considering the ladder diagram with n-rungs. If the parameters of the rungs are $x_1 \dots x_n$ and the parameters of the sides are y_i then (3.6.13) becomes

$$L_n(\beta) = \Gamma(-\beta) g^2 \left(\frac{-g^2}{16\pi^2}\right)^{n-1} (-1)^n$$
$$\times \int_0^\infty dx_1 \dots dx_n \Pi dy\, (x_1 \dots x_n)^\beta\, [C(x, y)]^{-2-\beta} e^{-J}. \tag{3.6.14}$$

This is defined for $\mathrm{Re}\,\beta > -1$ but becomes divergent at $\beta = -1$ because each of the x_i integrations then diverges at zero. The singularity corresponding to these divergent integrations can be exhibited by integrating by parts with $\operatorname{Re}\beta > -1$. This gives as equivalent to (3.6.14)

$$L_n(\beta) = \Gamma(-\beta)g^2 \left(\frac{-g^2}{16\pi^2}\right)^{n-1} \int_0^\infty dx_1 \dots dx_n \, \Pi dy$$

$$\times \frac{(x_1 \dots x_n)^{\beta+1}}{(\beta+1)^n} \frac{\partial^n}{\partial x_1 \dots \partial x_n} (C^{-2-\beta}e^{-J}). \quad (3.6.15)$$

The integral here is convergent for $\operatorname{Re}\beta > -2$ and so (3.6.15) provides a continuation below $\operatorname{Re}\beta = -1$. The multiple pole at $\beta = -1$ has been explicitly exhibited. The highest order singularity at $\beta = -1$ is the pole of order n and its residue is obtained by putting $\beta = -1$ elsewhere in the integral. The $x_1, \dots, x_n$ integrals can then be performed, with the effect of substituting $x_i = 0$. This is the leading approximation and the remaining integral over the y_i is easily identified with the coefficient already calculated in (3.3.14). We, however, are also interested in the lower order poles at $\beta = -1$ as we wish to do better than the leading order approximation. To calculate these coefficients we expand the β-dependence of the remainder of the integrand.† This gives for the coefficient of $(\beta+1)^{-p}$

$$L_n(p) = \Gamma(-\beta)g^2 \left(\frac{-g^2}{16\pi^2}\right)^{n-1} \int_0^\infty dx_1 \dots dx_n \, \Pi dy_i$$

$$\times \sum_{q=0}^{n-p} \frac{(\sum_j \log x_j)^{n-p-q}}{(n-p-q)!} \frac{\partial^n}{\partial x_1 \dots \partial x_n} \left[\frac{e^{-J}(-\log C)^q}{C \cdot q!}\right]. \quad (3.6.16)$$

In order to determine the structure of (3.6.16) we expand the first term in its integrand by the multinomial theorem to give a sum of terms of the form

$$\prod_{j=1}^n \frac{(\log x_j)^{s_j}}{s_j!}, \quad \sum_j s_j = n-p-q. \quad (3.6.17)$$

In this term any x_j for which $s_j = 0$ corresponds to an integration which can immediately be performed since the only x_j dependence is then inside the differential operator. Performing the integration has the effect of putting x_j zero inside the square brackets. Since this corresponds to contracting a rung of the ladder, and thus producing a contracted diagram which consists of two subdiagrams joined through a single vertex, the functions e^{-J} and C correspondingly factorise into two parts. When C factorises, the logarithm in the square brackets can also be expanded by the multinomial theorem.

† It is technically convenient not to expand the $\Gamma(-\beta)$ factor. If we sum *all* the terms it makes no difference whether $\Gamma(-\beta)$ is expanded or not.

These factorisations and expansions lead to a series of terms which can be summed. The details are in the appendix to this section. The final form obtained is

$$\sum_{n=1}^{\infty} \sum_{p=0}^{n} L_n(p) \cdot (\beta+1)^{-p} = \mathscr{G}(\beta, s) \frac{\Gamma(-\beta)}{\mathscr{F}(\beta, s)} \mathscr{G}(\beta, s), \quad (3.6.18)$$

where $\mathscr{F}$ and $\mathscr{G}$ are defined by (A 3.6.8, 9).

The Regge poles of the theory will be given by the roots of

$$\mathscr{F}(\beta, s) = 0. \quad (3.6.19)$$

If $\beta_0(s)$ is a root of this equation it gives a term in the amplitude calculated from (3.6.2) and (3.6.18) of the form

$$\sim \Gamma(-\beta_0(s)) \mathscr{G}^2(\beta_0(s), s)(-t)^{\beta_0(s)}. \quad (3.6.20)$$

The poles of $\Gamma(x)$ at the non-positive integers will give poles of (3.6.20) when β_0 takes non-negative integral values. Thus (3.6.20) exemplifies in a direct way the connection between poles in the s-channel at integral angular momentum and the high energy behaviour in the t-channel.

The form of the equation (3.6.19) resulting from (A 3.6.8) is very similar to that obtained by Cassandro, Cini, Jona-Lasinio & Sertorio (1963), for scattering by Yukawa potentials. In particular, it may be shown to give the Gribov–Pomeranchuk (1962 b) threshold behaviour near $\mathrm{Re}\,\beta = -\frac{1}{2}$, discussed in §3.2. To see this it is sufficient to consider the first approximation to (A 3.6.8), given by

$$\beta + 1 - \overline{F}_0(\beta, s) = 0. \quad (3.6.21)$$

The integral corresponding to (A 3.6.10) may be written in the more familiar Feynman form as

$$\overline{F}_0(\beta, s) = \frac{g^2}{16\pi^2} \Gamma(-\beta) \int_0^1 \frac{dx}{[\epsilon x(1-x) + (1-2x)^2]^{-\beta}}, \quad (3.6.22)$$

where for unit masses $\quad \epsilon = 4 - s. \quad (3.6.23)$

The form of (3.6.22) near $\epsilon = 0$ may be evaluated by elementary means and is found to be

$$\overline{F}_0 \sim \frac{g^2}{16\pi^2} \left[\frac{\Gamma(-\beta)}{2\beta+1} - \frac{\pi \Gamma(\frac{1}{2}) \Gamma(\beta + \frac{3}{2})}{2 \sin \pi(\beta + \frac{1}{2})} (\tfrac{1}{4}\epsilon)^{\beta + \frac{1}{2}} \right]. \quad (3.6.24)$$

The pole of (3.6.24) at $\beta = -\frac{1}{2}$ leads to an infinite number of solutions of the transcendental equation (3.6.21) which have $\mathrm{Re}\,\beta = -\frac{1}{2}$ as $\epsilon \to 0$ (see Cassandro *et al.* 1963).

The Mellin transform method has also been used by Swift (1965) to investigate the behaviour of ladder diagrams near $\beta = -2$. He finds a term representing the next-to-leading term in the expansion of the Legendre function associated with the $\beta = -1$ trajectory, together with another term corresponding to a new trajectory associated with $\beta = -2$.

Appendix to § 3.6

In this appendix we show how the expansions and factorisations discussed in the paragraph following (3.6.17) lead to terms which can be completely summed (Polkinghorne, 1964 a). Their effect is to produce expressions built up out of the following basic units:

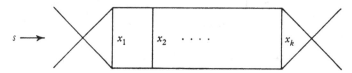

Fig. 3.6.1. A contracted interior subdiagram.

$$F_k(s_1, ..., s_k; \gamma) \equiv -\left(\frac{-g^2}{16\pi^2}\right)^{k+1} \int_0^\infty dx_1 ... dx_k \, \Pi dy_i$$

$$\times \prod_{j=1}^{k} \frac{(\log x_j)^{s_j}}{s_j!} \frac{\partial^k}{\partial x_1 ... \partial x_k} \left[\frac{(-\log C_k)^\gamma e^{-J_k}}{\gamma! \, C_k}\right], \quad (A\,3.6.1)$$

with $s_j > 0$ and C_k and J_k corresponding to Fig. 3.6.1;

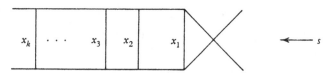

Fig. 3.6.2. A contracted end subdiagram.

$$G_k(s_1, ..., s_k; \gamma) \equiv -g\left(\frac{-g^2}{16\pi^2}\right) \int_0^\infty dx_1 ... dx_k \, \Pi dy$$

$$\times \prod_{j=1}^{k} \frac{(\log x_j)^{s_j}}{s_j!} \frac{\partial^k}{\partial x_1 ... \partial x_k} \left[\frac{(-\log \bar{C}_k)^\gamma e^{-\bar{J}_k}}{\gamma! \, \bar{C}_k}\right], \quad (A\,3.6.2)$$

with $s_j > 0$ and $\bar{C}_k$ and $\bar{J}_k$ corresponding to Fig. 3.6.2. It is formally convenient to allow k also to take the value zero in F_k to correspond to the simple bubble diagram contribution, $K(s)$. Similarly, we define $G_0 = g$. Both F_k and G_k are, of course, functions of s.

The terms that appear in the expansion are all of the form of a product of a G times a certain number of F's times another G. These terms are formed by contracting ladder diagrams of appropriate length. In this way a given set of F-factors will appear in a number of distinct ways which is just given by a multinomial coefficient corresponding to the number of distinct orders in which they can be arranged along a line. If the corresponding power of $(\beta+1)^{-1}$ were just given by the number of these F-factors then summation would be immediate. However, this is not the case. In fact the power of $(\beta+1)^{-1}$ associated with a given $G \cdot \Pi F \cdot G$ product is $1 + \Sigma n_i$ where the n associated with an F factor is

$$n = 1 - \Sigma(s_j - 1) - \gamma, \tag{A 3.6.3}$$

and the n associated with a G-factor is

$$n = -\Sigma(s_j - 1) - \gamma. \tag{A 3.6.4}$$

These equations make it convenient to define new functions

$$F'_k(\beta; s_1, ..., s_k; \gamma) \equiv \prod_{j=1}^{k} (1+\beta)^{s_j-1}(1+\beta)^\gamma F_k(s_1, ..., s_k; \gamma), \tag{A 3.6.5}$$

and G'_k similarly defined. Then the sum over all terms has the form

$$L = \sum_{f=0}^{\infty} \Big(\sum_{k, s_j, \gamma} G'_k \Big) \Big(\sum_{k, s_j, \gamma} F'_k \Big)^f \Big(\sum_{k, s_j, \gamma} G'_k \Big)(\beta+1)^{-f-1}. \tag{A 3.6.6}$$

This sums to give

$$L = \mathscr{G} \cdot \frac{1}{\mathscr{F}} \cdot \mathscr{G}, \tag{A 3.6.7}$$

where

$$\mathscr{F}(\beta, s) = \beta + 1 - \sum_{k=0}^{\infty} \bar{F}_k, \tag{A 3.6.8}$$

$$\mathscr{G}(\beta, s) = \sum_{k=0}^{\infty} \bar{F}_k. \tag{A 3.6.9}$$

The functions $\bar{F}_k$ and $\bar{G}_k$ are obtained by performing the sums over the s_j and γ to give

$$\bar{F}_k = -\left(\frac{-g^2}{16\pi^2}\right)^{k+1} \int_0^\infty dx_1 \dots dx_k \, \Pi dy$$

$$\times \prod_{j=1}^{k} \frac{(x_j^{\beta+1} - 1)}{\beta+1} \frac{\partial^k}{\partial x_1 \dots \partial x_k} [C_k^{-2-\beta} e^{-J_k}], \tag{A 3.6.10}$$

together with a similar definition of $\bar{G}_k$.

3.7   Pinch contributions and the Gribov–Pomeranchuk phenomenon

General ideas

In addition to the end-point contributions already considered there will be contributions to the high-energy behaviour from internal regions of the hypercontour where g is zero, which are trapped by a pinch as $t \to \infty$ (Polkinghorne, 1963d; Tiktopoulos, 1963b). The rules for the structure of D (§ 1.5) make it clear that in the physically relevant case of positive α's this can only happen for contributions from non-planar diagrams. Planar diagrams correspond to expressions for g which are just sums of products of α's, and when the α's are non-negative these can only vanish by some of the α's actually being zero. All diagrams with a third Mandelstam spectral function, ρ_{tu}, are non-planar and these are the diagrams which do not have an analogue in potential theory. Thus the study of pinch contributions is of particular interest in that they correspond to the new features of a fully relativistic crossing-symmetric theory which cannot be reproduced in a potential-theory model.

We have already encountered one such relativistic effect in the GP phenomenon of § 3.2. We recall that this predicts the existence of an essential singularity at $l = -1$ and that this is due to: (*a*) a non-zero third spectral function ρ_{tu}; (*b*) a region of two-particle unitarity in the s-channel. In this section we shall find the same phenomenon associated with a definite sequence of Feynman diagrams. First, however, it will be instructive to consider a model integral which illustrates the main features of pinch contributions. It is

$$I = \int_{x_1}^{x_2} dx \int_{y_1}^{y_2} dy \, \frac{1}{[xyt+d]^{n+2}}. \tag{3.7.1}$$

The integrations are easily performed to give

$$-\frac{1}{(n+1)t\,d^{n+1}} \log \left[\frac{(x_1 y_1 t+d)\,(x_2 y_2 t+d)}{(x_1 y_2 t+d)\,(x_2 y_1 t+d)} \right] + R(t), \tag{3.7.2}$$

where $R(t)$ is a rational function of t whose explicit form need not be given.

As $t \to \infty$ the logarithm in (3.7.2) tends to $\log 1$. If the principal branch of the logarithm is chosen, then cancellations between the two terms in (3.7.2) produce a net asymptotic behaviour of t^{-n-2} as $t \to \infty$, provided that $x_1, \ldots, y_2$ are all non-zero. However, if we choose the branch of the logarithm in which $\log 1 = 2m\pi i$ (m an integer) then

$$I \sim -\frac{2m\pi i}{(n+1)\,d^{n+1}}\,t^{-1}. \tag{3.7.3}$$

If $x_1, \ldots, y_2$ are all positive the principal branch is the correct choice. However, if $x_1, y_1 < 0$; $x_2, y_2 > 0$; then as $t \to \infty + i\epsilon$, the top two factors in the logarithm tend to $+\infty + i\epsilon$, while the bottom two factors each tend to $-\infty - i\epsilon$. The correct branch is then given by $m = -1$ and $I \sim t^{-1}$.

Behaviour which depends upon the branch of the functions considered is typical of pinch mechanisms, since on some sheets the contour will be actually trapped by the pinch, whilst on others it will not. It is easy to see how (3.7.3) results from a pinch. In order to be able to discuss $t = \infty$, it is convenient to make the change of variable

$$t = t_0 z^{-1}, \tag{3.7.4}$$

and consider the limit $z \to 0$, at fixed t_0. A pinch will occur when

$$\left.\begin{aligned} xyt_0 + dz &= 0, \\ yt_0 &= 0, \\ xt_0 &= 0. \end{aligned}\right\} \tag{3.7.5}$$

These equations have a solution $x = y = 0$ and $z = 0$. Thus if the contours of integration pass through the origin they will be trapped as $t \to \infty$, and give the behaviour (3.7.3).

Iterated crosses

The simple result (3.7.3) provides an analytic tool which can be put to immediate use. The diagram of Fig. 3.7.1 is the least complicated one which might be expected to show some trace of the GP phenomenon. It contains the first two-particle iteration in the s-channel of the cross-diagram, the simplest diagram with a ρ_{tu} spectral function. We have already seen in §3.4, that a single cross has three independent scalings of length 2. Thus Fig. 3.7.1 will have six scalings of length 2 giving it an end-point contribution of $t^{-2}\,(\log t)^5$. However the GP phenomenon is first expected at $l = -1$, not $l = -2$. It must therefore come from a pinch contribution. The coefficient of t in the Feynman denominator is of the form

$$xy = (\alpha_1 \alpha_3 - \alpha_2 \alpha_4)\,(\alpha_1' \alpha_3' - \alpha_2' \alpha_4'). \tag{3.7.6}$$

The integration over positive α's will include points in its interior which correspond to $x = y = 0$. In the neighbourhood of these points the integral may be approximated by an expression of the form (3.7.1) in order to calculate the leading behaviour. Thus an asymptotic behaviour of t^{-1} is obtained.

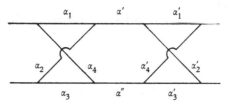

Fig. 3.7.1. A diagram associated with the GP phenomenon.

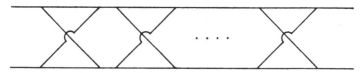

Fig. 3.7.2. A typical member of a set of diagrams giving an essential singularity at $l = -1$.

The singularity at $l = -1$ thus obtained from Fig. 3.7.1 is just a simple pole. In order to see that in reality an essential singularity exists there, it is necessary to consider the complete two particle iteration, that is the set of diagrams containing all numbers of crosses inserted (Fig. 3.7.2). The proof that this set leads to essential singularity was first given by Contogouris (1965) and Kaschlun & Zoellner (1965) using the Bethe–Salpeter equation. Here we use a somewhat different approach.

The coefficient of t in a diagram like Fig. 3.7.2 is of the form $\prod_i x_i$, where each x_i has the form

$$x_i = \alpha_1^i \alpha_3^i - \alpha_2^i \alpha_4^i, \tag{3.7.7}$$

the α^i being the Feynman parameters of the ith cross. Consequently our model integral must be a generalisation of (3.7.1) of the form

$$I_n = \prod_{i=1}^{r} \int_{x_i'}^{x_i''} dx_i \frac{1}{[\prod_i x_i t + d]^{n+2}} \quad (x_i' < 0, \, x_i'' > 0, \, i = 1, \dots r). \tag{3.7.8}$$

In order to evaluate the leading contribution from the pinch at $x_i = 0$ it is convenient to express the pinch contribution as a difference of end-point contributions.

Each integration is decomposed into two parts

$$\int_{x_i'}^{x_i''} dx_i \to \int_0^{x_i''} dx_i + \int_{x_i'}^0 dx_i, \tag{3.7.9}$$

and the integrations over negative ranges of x_i are transformed into integrations over positive ranges by the substitution $x_i \to -x_i$. All terms in the sum of integrals obtained via (3.7.9) which had an even number of negative range integrations will now have an identical form and so give identical leading behaviour. There are 2^{r-1} such integrals and each gives an end-point contribution

$$\frac{1}{(n+1)\, d^{n+1}}\, t^{-1}\, \frac{(\log t)^{r-1}}{(r-1)!}. \tag{3.7.10}$$

Similarly, the 2^{r-1} integrals which had an odd number of negative range integrations also have identical forms and give an asymptotic behaviour which is just obtained from (3.7.10) by the substitution $t \to -t$; $\log t \to \log t - \pi i$. Thus the leading behaviour of the sum is just

$$\frac{2^{r-1}\pi i}{(n+1)\, d^{n+1}}\, t^{-1}\, \frac{(\log t)^{r-2}}{(r-2)!}. \tag{3.7.11}$$

Notice that if d depended on the x_i then we would get a similar result except that all the x_i must be put zero in the d which appears in (3.7.11).

When this result is applied to Fig. 3.7.2 with r-crosses we immediately obtain for the leading behaviour a singularity in the l-plane of the form

$$G_0(2r-2)!\, \frac{\pi i\, G^{r-1}}{(l+1)^{r-1}} \int d\alpha\, \frac{\prod_i \delta(x_i)\, \delta(\Sigma\alpha - 1)\, C^{2r-2}}{D^{2r-1}}, \tag{3.7.12}$$

where C and D are the Feynman functions of the diagram, x_i is defined by (3.7.7), and G_0 and G are coupling constant factors.

The expression (3.7.12) does not factorise. We do not, indeed, expect that it should, for then it would give a Regge pole and not an essential singularity. It may, however, be cast into a more perspicuous form by scaling over the parameters of each cross. To avoid the purely formal complications of the general case we shall in fact be content to illustrate the procedure by applying it to Fig. 3.7.1. If λ and λ' are the scaling parameters of the two crosses then (3.7.12) becomes

$$G_0 \cdot 2 \cdot \frac{\pi i G}{(l+1)} \int d\lambda\, d\lambda'\, d\bar{\alpha}_i\, d\bar{\alpha}_i'\, d\alpha'\, d\alpha''\, \delta(\bar{x})\, \delta(\bar{x}')\, \delta(\Sigma\bar{\alpha}_i - 1)$$

$$\times \delta(\Sigma\bar{\alpha}_i' - 1)\, \delta(\lambda_1 + \lambda_2 + \alpha' + \alpha'' - 1)\frac{[\bar{C}]^2}{[\bar{D}]^2}, \tag{3.7.13}$$

where
$$\begin{aligned} \bar{x} &= \lambda^{-2}x = (\bar{\alpha}_1\bar{\alpha}_3 - \bar{\alpha}_2\bar{\alpha}_4), \\ \bar{x}' &= (\lambda')^{-2}x' = (\bar{\alpha}_1'\bar{\alpha}_3' - \bar{\alpha}_2'\bar{\alpha}_4'), \\ \bar{C} &= (\lambda\lambda')^{-1}C; \quad \bar{D} = (\lambda\lambda')^{-1}D. \end{aligned} \right\} \tag{3.7.14}$$

The expression (3.7.13) may be rewritten

$$\frac{G_0 G^{-1}\pi i}{(l+1)} \int d^2k \, \frac{f(p_1, p_2; \, k, -k+p)f(k, -k+p; \, p_3, p_4)}{[k^2 - m^2][(k-p)^2 - m^2]}, \tag{3.7.15}$$

where the momenta are defined by Fig. 3.7.3, the integration is over a two-dimensional loop momentum k, and f is a function associated with a single cross diagram and given by

$$f = \int_0^1 d\alpha_1 \dots d\alpha_4 \, \frac{\delta(\Sigma\alpha - 1)\,\delta(\alpha_1\alpha_3 - \alpha_2\alpha_4)}{d(s; \, \alpha)}; \tag{3.7.16}$$

d is the Feynman denominator associated with the single cross diagram (Fig. 3.4.8). Notice that because of the second δ-function in (3.7.16) d is effectively independent of t.

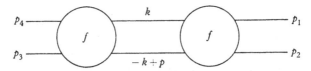

Fig. 3.7.3. A symbolic representation of the coefficient of the GP singularity associated with Fig. 3.7.1.

The proof of the equivalence of (3.7.13) and (3.7.15) is most easily given by considering the identities which must exist between the functions C, D and d resulting from the fact that Fig. 3.7.1 can be thought of as being built up out of two single crosses and an integral over the loop joining them.

The general result is just the iteration of (3.7.15), each additional cross providing a factor $f/(l+1)$ which is joined on by a two-dimensional loop momentum. Symbolically we get for the sum

$$\boxed{F} = \sum_n \boxed{f}\boxed{f} \dots \boxed{f} \, (l+1)^{-n+1}, \tag{3.7.17}$$
$$[n\text{-factors}]$$

which may be rewritten in integral equation form

$$\boxed{F} = \boxed{f} + \frac{1}{(l+1)} \boxed{f}\boxed{F}. \tag{3.7.18}$$

A trivial manipulation enables us to write the kernel in symmetric form. It is bounded, except at $l = -1$. Theorems on the limit points of the eigenvalues of such integral equations (Courant & Hilbert, 1953), establish a condensation of poles at $(l+1)^{-1} = \infty$, i.e. $l = -1$.

Heuristically we can see this as follows. If we expand f in terms of its eigenvectors:

$$f = \sum_\alpha \lambda_\alpha |\alpha\rangle \langle\alpha|, \tag{3.7.19}$$

the boundedness of the trace implies $\lambda_\alpha \to 0$ as $\alpha \to \infty$. However, F is also now diagonal and can be written

$$F = \sum_\alpha |\alpha\rangle \frac{\lambda_\alpha}{1 - \lambda_\alpha/(l+1)} \langle\alpha|, \tag{3.7.20}$$

giving a sequence of poles $l = -1 + \lambda_\alpha$ which accumulate at $l = -1$.

3.8   Regge cuts

Introduction

The proposal that in a relativistic theory there might be cuts in the angular momentum plane was first made by Amati, Fubini & Stanghellini (1962). The diagram they considered was essentially Fig. 3.8.1. The discontinuity of this diagram across its two-particle normal threshold in the t-channel is given by (Drummond 1963),

$$\int \frac{ds_1 \, ds_2}{\Delta^{\frac{1}{2}}(s, t; \, s_1, s_2, m_1^2, m_2^2)} f_1(s_1, t) f_2(s_2, t), \tag{3.8.1}$$

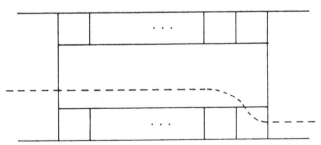

Fig. 3.8.1. The *AFS* diagram.

where f_1 and f_2 are scattering amplitudes corresponding to the two Regge poles in the two halves of the figure and Δ is proportional to the left-hand side of (2.4.10). Notice that such an expression is only possible in a crossing-symmetric theory, where there must be unitarity in the t-channel as well as in the s-channel. Since

$$\Delta^{\frac{1}{2}} \sim t \,.\, \lambda^{\frac{1}{2}}(s, s_1, s_2) \quad \text{as} \quad t \to \infty, \tag{3.8.2}$$

where

$$\lambda(s, s_1, s_2) = s^2 + s_1^2 + s_2^2 - 2ss_1 - 2ss_2 - 2s_1 s_2, \tag{3.8.3}$$

the high t behaviour of (3.8.1) will be given by

$$\int \frac{ds_1 \, ds_2 \, \beta(s_1) \, \beta(s_2)}{\lambda^{\frac{1}{2}}(s, s_1, s_2)} \, t^{\alpha(s_1) + \alpha(s_2) - 1}. \tag{3.8.4}$$

Since (3.8.4) gives a continuous range of values of the exponent of t, it must correspond to a cut in the complex angular momentum plane (cf. (3.1.8)). The location of the singularity trailing the cut will depend on s. It is therefore called a moving cut.

Nevertheless, the complete contribution associated with Fig. 3.8.1 cannot have this cut in the physical limit (Polkinghorne, 1963a; Mandelstam, 1963a) since it is a planar diagram whose behaviour is completely given by end-point contributions. However many rungs there may be in the two ladders, the diagram never has more than two d-lines of length 3 and so its asymptotic behaviour is simply $t^{-3} \log t$. It can be shown (Polkinghorne, 1962c) that the discontinuities of Fig. 3.8.1 across its higher normal thresholds (such as the three particle threshold given by the dotted line in the figure) also have cuts of identical form to (3.8.4). In the complete contribution all these cut terms must just cancel. The cut (3.8.4) is in fact present only on unphysical sheets (Polkinghorne, 1963d).

Physical sheet moving cuts

It is clear that we can only expect moving cuts on the physical sheet from non-planar diagrams. Mandelstam (1963b) has given arguments, to which we shall return later, that the diagram of Fig. 3.8.2 should be expected to give a cut on the physical sheet. In order to simplify the discussion one of the Regge poles has been replaced by an elementary pole. The occurrence of the cut for Fig. 3.8.2 can be shown explicitly (Polkinghorne, 1963d).

The coefficient of t in the diagram has the form

$$g = xyC(L_1 \dots L_n) + P_1 x + P_2 y + Q, \tag{3.8.5}$$

where

$$\begin{rcases} x = \alpha_1 \alpha_3 - \alpha_2 \alpha_4, \\ y = \alpha_1' \alpha_3' - \alpha_2' \alpha_4', \end{rcases} \tag{3.8.6}$$

and $C(L_i \ldots L_j)$ is the C-function associated with the subdiagram formed by combining the loops $L_i \ldots L_j$.† The remaining coefficients in (3.8.5) are given by

$$P_1 = \sum_{i=1}^{n} C(L_1 \ldots L_{i-1})(\gamma_i \alpha_1' - \gamma_{n+i}\alpha_4') \prod_{k>i} \beta_k, \qquad (3.8.7)$$

$$P_2 = \sum_{i=1}^{n} (\gamma_i \alpha_1 - \gamma_{n+i}\alpha_4) C(L_{i+1} \ldots L_n) \prod_{k \leqslant i} \beta_k. \qquad (3.8.8)$$

$$Q = \beta_1 \ldots \beta_{n+1}[\alpha_1 \alpha_1' C(L) + \alpha_4 \alpha_4' C(L') + \alpha_1 \alpha_4' \delta + \alpha_1' \alpha_4 \delta]$$
$$+ \sum_{i<j} \prod_{k \leqslant i} \beta_k C(L_{i+1} \ldots L_{j-1}) \prod_{l>j} \beta_l(\alpha_1 \gamma_i - \alpha_4 \gamma_{n+i})(\alpha_1' \gamma_j - \alpha_4' \gamma_{n+j}).$$
$$(3.8.9)$$

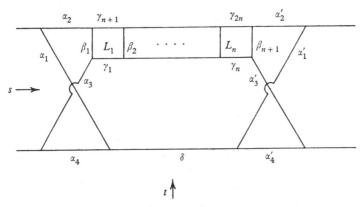

Fig. 3.8.2. A diagram giving a Regge cut on the physical sheet.

The loop L is formed by the lines corresponding to $\alpha_4 \alpha_3 \gamma_1 \ldots \gamma_n \alpha_3' \alpha_4' \delta$, and the loop L' by the lines corresponding to $\alpha_1 \alpha_2 \gamma_{n+1} \ldots \gamma_{2n} \alpha_2' \alpha_1' \delta$.

In order to extract the pinch behaviour of the diagram a generalisation of the simple standard form (3.7.1) is required. We consider the integral

$$I' = \int_{x'}^{x''} dx \int_{y'}^{y''} dy \int_{0}^{\epsilon} d\beta_1 \ldots d\beta_{n+1} \frac{1}{[(xy + \beta_1 \ldots \beta_{n+1})t + d]^{p+2}};$$
$$x',y' < 0, \ x'',y'' > 0. \quad (3.8.10)$$

A simple computation shows that its leading asymptotic behaviour is

$$\frac{2\pi i}{(p+1)pd^p} t^{-2} \frac{(\log t)^n}{n!}. \qquad (3.8.11)$$

† It will be convenient in what follows to define $C = 1$ if $j < i$.

This results from the combination of a pinch in x and y with end-points in the β's.

To apply (3.8.10) to the Feynman integral of Fig. 3.8.2 we rewrite (3.8.5) in the form

$$g = C(x + P_2 C^{-1})(y + P_1 C^{-1}) + Q - (P_1 P_2 C^{-1}). \qquad (3.8.12)$$

Examination of $Q - (P_1 P_2 C^{-1})$ shows that it vanishes if any one of the β's is set equal to zero. In the first instance, therefore, we look at the effect of a pinch at $x = -P_2 C^{-1}$, $y = -P_1 C^{-1}$ together with end-points at $\beta_i = 0$ $(i = 1, ..., n+1)$. The pinch produces factors in the integrand

$$\delta(x + P_2 C^{-1})\,\delta(y + P_1 C^{-1}). \qquad (3.8.13)$$

These enable the α_4 and α_4' integrations to be performed with the effect of producing the substitutions

$$\left. \begin{aligned} \alpha_4 &= \alpha_1 \alpha_3 \alpha_2^{-1} + O(\beta), \\ \alpha_4' &= \alpha_1' \alpha_3' (\alpha_2')^{-1} + O(\beta), \end{aligned} \right\} \qquad (3.8.14)$$

together with the multiplication of the integrand by $(\alpha_2 \alpha_2')^{-1} + O(\beta)$. When the β's are set equal to zero after scaling, the substitutions (3.8.14) have the effect of making $Q - (P_1 P_2 C^{-1})$ also vanish if α_1 or α_1' is set equal to zero so that the integrations over these variables also contribute to the asymptotic behaviour. The final expression is therefore

$$2\pi g^4 \left\{ i \left(\frac{-g^2}{16\pi^2} \right) (n+1)! \int_0^1 d\alpha_2\, d\alpha_3\, d\alpha_2'\, d\alpha_3'\, d\gamma_i\, d\delta \right.$$

$$\left. \times \frac{\delta(\alpha_2 + ... + \alpha_3' + \delta + \Sigma\gamma - 1)\,[c(\alpha, \gamma, \delta)]^{n+2}}{f(\alpha, \gamma, \delta)\,[d(s; \alpha, \gamma, \delta)]^{n+2}} \right\} \frac{t^{-2}(\log t)^{n+2}}{(n+2)!}, \qquad (3.8.15)$$

where c and d refer to the contracted diagram in Fig. 3.8.3, and the function f is defined by

$$f = \lim_{\substack{\alpha_1 \to 0 \\ \alpha_1' \to 0 \\ \beta_i \to 0}} \left[\frac{\alpha_2 \alpha_2'(QC - P_1 P_2)}{\alpha_1 \alpha_1' \beta_1 ... \beta_{n+1}} \right], \qquad (3.8.16)$$

where the substitutions (3.8.14) have been made in $QC - P_1 P_2$. From equations (3.8.7–9) f may be evaluated and a somewhat tedious enumeration of the terms occurring shows that

$$f(\alpha, \gamma, \delta) = c(\alpha, \gamma, \delta). \qquad (3.8.17)$$

This identification has the effect of making the integral in curly brackets in (3.8.15) just the correct Feynman integral associated with

Fig. 3.8.3 evaluated with two-dimensional loop momenta. If this two-dimensional integral is written in terms of invariants (Drummond, 1963), (3.8.15) appears in the form

$$2\pi g^4 \int\limits_{\lambda \leqslant 0} \frac{ds_1 ds_2}{\lambda^{\frac{1}{2}}(s, s_1, s_2)} \frac{[K(s_1)]^{n+2}}{s_2 - m^2} t^{-2} \frac{(\log t)^{n+2}}{(n+2)!}, \qquad (3.8.18)$$

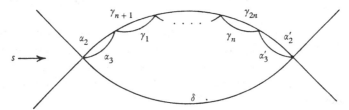

Fig. 3.8.3. The contracted diagram associated with Fig. 3.8.2.

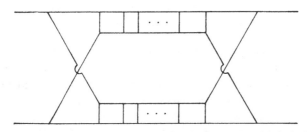

Fig. 3.8.4. Another diagram giving a Regge cut on the physical sheet.

where λ is defined by (3.8.3) and $K(s)$ is the function associated with the ladder trajectory given by (3.3.9). When (3.8.18) is summed over n it gives a cut contribution of the form

$$2\pi g^4 \int\limits_{\lambda \leqslant 0} \frac{ds_1 ds_2}{\lambda^{\frac{1}{2}}(s, s_1, s_2)} \frac{t^{\alpha(s_1)-1}}{s_2 - m_2^2}. \qquad (3.8.19)$$

The possibility of scaling α_1 and α_1' is a peculiarity of the diagram, Fig. 3.8.2, with its single ladder insertion. If the diagram of Fig. 3.8.4 is considered, its leading singularity is associated with a pinch due to the two crosses together with end-points corresponding to putting a parameter from each ladder equal to zero. In evaluating the latter effect care must be taken to count all the independent scaling procedures as explained at the end of § 3.4 (cf. also the discussion in § 3.10). The resulting asymptotic behaviour is found to be

$$\int \frac{[f(s, s_1, s_2)]^2 ds_1 ds_2}{\lambda^{\frac{1}{2}}(s, s_1, s_2)} [K(s_1)]^m [K(s_2)]^n \frac{(m+n)!}{m! \, n!} t^{-3} \frac{(\log t)^{m+n}}{(m+n)!}, \qquad (3.8.20)$$

where m and n are the number of loops in the two ladders respectively, and f is the function associated with the cross-diagram and defined by (3.7.16). When this is summed over m and n it gives an expression

$$\int \frac{[f(s, s_1, s_2)]^2 \, ds_1 \, ds_2}{\lambda^{\frac{1}{2}}(s, s_1, s_2)} t^{\alpha(s_1) + \alpha(s_2) - 1}, \qquad (3.8.21)$$

which corresponds to a cut with the same trajectory as that given by (3.8.4).

It is now possible to see the connection between the expressions we have obtained and the way in which Mandelstam (1963b) has discussed Regge cuts. We do so heuristically since a full discussion involves technical complexities which cloud the simplicity of the basic idea.

In terms of l (3.8.21) corresponds to a singularity

$$\int \frac{f^2 \, ds_1 \, ds_2}{\lambda^{\frac{1}{2}}} \frac{1}{l - \alpha(s_1) - \alpha(s_2) + 1}. \qquad (3.8.22)$$

This is, of course, just the expression given by the leading order approximation. If we used the methods of § 3.6, suitably generalised, to take further lower order terms into account, it would have two effects. One would be to produce more complicated expressions for the trajectories. This is not specially interesting for our present purpose which uses perturbation theory as a model, since we do not want to take very specific details seriously. The second effect would be to produce terms, analogous to $\Gamma(-\alpha)$ in § 3.6, which give poles when $\alpha(s)$ takes non-negative integral values. This is an important feature which is clearly more general than the specific model. The poles at $\alpha(s) = 0$ give a singularity in (3.8.22) at $l = 1$, which is connected with the singularity (3.7.15) associated with Fig. 3.7.1, the figure obtained from Fig. 3.8.4 by replacing the two ladders by similar particles. Mandelstam (1963b) was able to show that the cut (3.8.22) switches on and off the corresponding GP singularity. He used this to argue that the cut could not be cancelled by other contributions. The lack of cancellation is confirmed in perturbation theory by the fact that other possible pinch contributions correspond to more elaborate insertions involving higher powers of the coupling constant.

Consequences

The existence of cuts considerably complicates the phenomenology of high-energy scattering. The exact position of the singularity trailing the cut is given by the maximum value of $\alpha(s_1) + \alpha(s_2) - 1$ in the region of integration. This depends on the nature of the trajectory function. If we make the simple assumption that $\alpha(s)$ depends linearly on s then we obtain a singularity, given by $s_1 = s_2$ on the boundary $\lambda = 0$, that is $s_1 = s_2 = (1/4)\,s$, and so

$$l = 2\alpha\left(\frac{s}{4}\right) - 1. \tag{3.8.23}$$

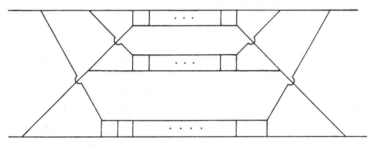

Fig. 3.8.5. A diagram compounding a cut and a pole.

If the constancy of high energy cross-sections (cf. § 3.2) is attributed to a Pomeranchukon whose trajectory passes through $l = 1$ when $s = 0$, then (3.8.23) implies that the cut also passes through $l = 1$ when $s = 0$. Moreover, instead of compounding two Regge poles to form a cut we can compound any pair of singularities, for example a cut and a pole, as in Fig. 3.8.5, or a pair of cuts. In this way a sequence of singularities is generated at positions

$$l = n\alpha\left(\frac{s}{n^2}\right) - (n-1), \quad n = 1, 2, 3.... \tag{3.8.24}$$

All pass through $l = 1$ when $s = 0$.

An attempt to determine the form of the diffraction peak from the superposition of the effect of these and other singularities, which are near $l = 1$ when s is near 0, has been made by Gribov, Pomeranchuk & Ter-Martirosyan (1965). Their analysis is based on a heuristic expression for the total discontinuity around the cut formed by combining two Regge poles. This has the form

$$\int \frac{ds_1 ds_2}{\lambda^{\frac{1}{2}}(s, s_1, s_2)} F^{(1)}(s, s_1, s_2; \alpha) F^{(2)}(s, s_1, s_2; \alpha)\, \delta(l - \alpha(s_1) - \alpha(s_2) + 1),$$

$$\tag{3.8.25}$$

where $F^{(1)}$ and $F^{(2)}$ are the amplitudes for the emission and absorption of the two Reggeons and the superscripts indicate that they are to be evaluated on opposite sides of the cut. This relation has a persuasive similarity to unitarity and its general form has been confirmed by a discussion of the Froissart–Gribov continuation (Polkinghorne, 1965b). However, a perturbation theory model discussed by Swift (1965) shows that the F's have properties rather different from those which simple intuition might have postulated.

The sequence of singularities (3.8.24) has a further disagreeable consequence. The Pomeranchukon must continue above $l = 1$ when $s > 0$. Let us suppose that for $s = s_0$, $\alpha(s) = 1 + \epsilon$.

For any finite s, $(s/n^2) \to 0$ as $n \to \infty$, and since $\alpha(s) \sim 1 + \alpha'(0)s$ for s sufficiently small, we see that the singularities (3.8.24) cannot extend more than a finite distance to the right for any fixed s. However, for $s = n^2 s_0$ the nth singularity is at $l = 1 + n\epsilon$. Thus as $s \to \infty$ there is no uniform bound limiting the extension of singularities to the right in the l-plane. Such a bound is necessary if double dispersion relations with a finite number of subtractions are to be possible. This bar to the existence of the Mandelstam representation was pointed out by Mandelstam (1963b).

3.9 Particles with spin: Reggeisation†

General ideas

It is clear that in considering complex angular momentum there will be differences between the behaviours of spinless particles and of particles with spin. The most obvious differences will simply be associated with the possibility of adding the spin angular momentum to the orbital angular momentum. However we shall also encounter some unexpected and highly interesting features.

In principle, since the effect of spin is simply to make the numerator function $N(\alpha)$ in (3.3.3) different from unity without altering the structure of the denominator function, the methods we have developed for the spinless case are immediately applicable to the more general case of particles with spin. In practice, however, the resulting expressions are extremely complicated. As a result all that has proved possible so far has been to investigate low-order diagrams without giving complete summations corresponding to infinite sequences of diagrams. Even for the diagrams which have been considered the

---

† In this section, in contrast with elsewhere in this book, we need a number of simple results on the propagators and interactions of particles of spins $\frac{1}{2}$ and 1. A convenient account may be found in the book by Umezawa (1956).

details are very involved. For a complete account of them reference must be made to the original papers. In this section we shall be content to survey the main ideas, methods and results.

The numerator factors are due to the presence of terms $\gamma q + m$, or $g_{\mu\nu} - q_\mu q_\nu / m^2$, associated with the propagators of particles of spins $\frac{1}{2}$ and 1 respectively, together with possible momentum factors at the vertices if there are derivative terms in the interactions. The presence of these factors affects the asymptotic behaviour in three principal ways:

(i) They may cause explicit powers of t, the asymptotic variable, to appear in the numerator. These powers may arise either directly from external momenta present in the q's or they may be due to the displacements of the origin of the internal momenta (see § 1.5) which are necessary for symmetric integration over the loop momenta.

This effect is of great importance. Since the denominator functions in the case of particles with spin have the same structure as those obtained in the spinless case they can by themselves give an asymptotic behaviour of at most $t^{-1} (\log t)^n$. However, the numerator factors can then transform this into $(\log t)^n$; $t (\log t)^n$; etc. In the complex angular momentum plane this means that singularities which in the spinless amplitudes are associated with $l = -1$ can be shifted to the right to $l = 0$, $l = 1$, etc. This translational effect of spin was first noted in a general way by Azimov (1963). It corresponds simply to the addition of angular momenta.

(ii) The internal loop momenta factors in the numerator will produce a decrease in the power of the denominator obtained after symmetric integration compared with the case where they are absent. If this power becomes equal to, or less than, the length of the d-lines then an enhanced asymptotic behaviour is obtained (see the remarks following equation (3.4.9)).

(iii) The class of singular configurations may be enlarged by the effect of numerator factors. We recall that the singular configurations discussed in § 3.4 were produced by adding one or two lines to a d-line set to form a closed loop. The limit of two extra lines arose because this represented the difference in the powers of C and D.

If there is a factor $k_i.k_j$ in the numerator, where k_i and k_j are the loop momenta of the ith and jth loops respectively, then, instead of producing (3.4.11), symmetric integration gives (Chisholm, 1952):

$$\frac{[\text{Adj } A]_{ij}}{C} \frac{C^{p-3}}{D^{p-1}}, \qquad (3.9.1)$$

where, as usual, A is the matrix of the quadratic form of the k's appearing in the denominator. If $i = j$ and we scale with ρ round the ith loop (3.9.1) gives a net factor of ρ^{-3}. Thus (3.9.1) with $i = j$ would have a singular configuration if three lines could be added to a d-line to complete the ith loop.

Reggeisation

The most interesting phenomenon which has been discovered for particles with spin is the Reggeisation effect suggested by Gell-Mann & Goldberger (1962; see also Gell-Mann, Goldberger, Low & Zachariasen, 1963; Gell-Mann, Goldberger, Low, Marx & Zachariasen, 1964; Gell-Mann, Goldberger, Low, Singh & Zachariasen, 1964; Polking-horne, 1964*b*).

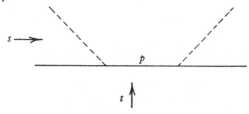

Fig. 3.9.1. The Born approximation diagram.

The theory considered is that of particles of spin $\frac{1}{2}$ (which we shall call nucleons) in interaction with neutral vector mesons through a conserved current. Because of the conserved current the $q_\mu q_\nu$ term in the vector meson propagator may be neglected and the theory is free from the unlimited divergence problems which otherwise afflict vector particles. There are still some divergences but the theory is re-normalizable (Schweber, 1961).

The simplest diagram is the Born approximation term of Fig. 3.9.1 corresponding to the contribution

$$\Gamma \frac{1}{\gamma p - m} \Gamma, \tag{3.9.2}$$

where Γ is the coupling to the external particles. It is in fact possible to consider the external particles as being pseudoscalar mesons: the essential role of the vector particles will be as internal particles generating a Regge pole from higher order diagrams. In this case[†]

$$\Gamma = \gamma_5.$$

[†] If the external particles are also taken as vector particles then a particular choice of gauge for Γ simplifies the calculation (see Gell-Mann, Goldberger, Low, Marx & Zachariasen 1964).

The asymptotic behaviour of (3.9.2) as $t \to \infty$ is constant, i.e. t^0 behaviour. In a theory with *spinless* particles this behaviour would make the term quite isolated as all other diagrams can at most behave like $t^{-1} (\log t)^n$ so that a t^0 term cannot be combined with anything else. This corresponds to the non-analytic behaviour of elementary particles in such a theory. A single particle pole of fixed angular momentum will only have an effect in the appropriate partial wave in which it can occur as an intermediate state. This is expressed by saying that it corresponds to a non-analytic behaviour in the Regge plane, such as that given by δ_{l0} if it only affects the s-state.

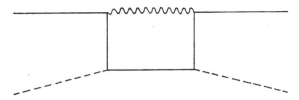

Fig. 3.9.2. A diagram to combine with Fig. 3.9.1.

However, in a theory with vector mesons we expect that the translational effect of spin will produce terms behaving like $(\log t)^n$ which might then combine with t^0 to give an entirely new type of behaviour. Gell-Mann & Goldberger (1962) suggested that this would happen in such a way that the nucleon actually lay on a Regge trajectory which had to pass through the nucleon pole when

$$ s \equiv p^2 = m^2. $$

In such a theory the nucleon then corresponds to a Regge pole and not to a non-analytic term, so the effect is called Reggeisation.

In order to find out if this can indeed happen the first diagram one must consider is that shown in Fig. 3.9.2, where the wavy line denotes the vector particle. One immediate complication is that in fact the appropriate asymptotic variable for Fig. 3.9.2 is clearly u ($\sim -t$ at fixed s). We shall see how to deal with this shortly. The leading asymptotic behaviour of Fig. 3.9.2 is straightforward to calculate and has the form

$$ \Gamma f(s) \, \Gamma \log u, \tag{3.9.3} $$

where $f(s)$ is a known function, involving γ-matrices, whose precise definition is not very interesting.

If one goes on to consider higher ladders formed out of nucleons and vector mesons then their asymptotic variables will alternate between

t and u according to whether they have an odd or even number of rungs. Such behaviour cannot correspond to a single Regge pole, but it must be associated with a pair of Regge poles, one of trajectory α and even signature, the other of trajectory $-\alpha$ and odd signature. Together they correspond to

$$\frac{\Gamma \cdot \frac{1}{2}[(t^{\alpha(s)} + u^{\alpha(s)}) + (t^{-\alpha(s)} - u^{-\alpha(s)})]\,\Gamma}{\gamma p - m}, \qquad (3.9.4)$$

which has an expansion in terms of $\log t$ and $\log u$ of the form

$$\frac{\Gamma \cdot (1 + \alpha(s)\log u + \alpha^2(s)\,(\log t)^2/2 + \dots)\,\Gamma}{\gamma p - m}. \qquad (3.9.5)$$

If we identify (3.9.2) and (3.9.3) with the first two terms of the series this tells us that
$$\alpha(s) = (\gamma p - m)f(s). \qquad (3.9.6)$$

The presence of the $(\gamma p - m)$ factor constrains $\alpha(s)$ to pass through zero when $s = p^2 = m^2$ for all values of the coupling constant. Thus the nucleon always lies on the Regge trajectory.

So far we have just conjectured that (3.9.4) represents the effect of summing up all relevant contributions. It is desirable to test this by at least looking for the next term in the series (3.9.5). The obvious diagram to consider is Fig. 3.9.3a. However, it turns out that by itself this diagram has an asymptotic behaviour $\sim (\log t)^3$. It is also necessary to consider the contributions from the crossed line diagrams, Figs. 3.9.3b, c. After many cancellations (Polkinghorne, 1964b)† the net contribution from all the diagrams of Fig. 3.9.3 is found to be

$$\Gamma f(s)\,(\gamma p - m)f(s)\,\Gamma \frac{(\log t)^2}{2}, \qquad (3.9.7)$$

as (3.9.5) and (3.9.6) together require.

Further diagrams

The translational effects of spin make it necessary to consider more diagrams than just the ladders and their associated crossed line diagram (Polkinghorne, 1964b). For example, the diagram of Fig. 3.9.4, which in the spinless case would be associated with a pair of poles near $l = -2$, can have an effect at $l = 0$ because of the presence

---

† *Note added in proof.* A term was omitted. The cancellation requires also the renormalization diagrams associated with Fig. 3.9.2 (Cheng & Wu, 1965).

of two vector particles in the intermediate state. It gives a contribution of the form

$$\Gamma f'(s)\,\Gamma \log u, \tag{3.9.8}$$

with $f'(s)$ a known function. There are two possible interpretations of (3.9.8). Either it represents a higher order term in the coupling constant

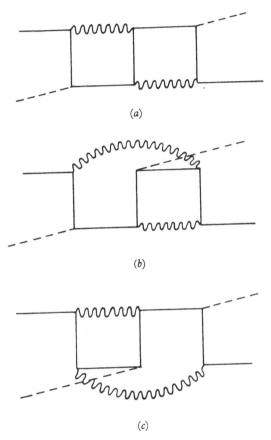

(a)

(b)

(c)

Fig. 3.9.3. Diagram related to Reggeisation in sixth order.

in the expansion of the trajectory $\alpha(s)$ that we have already considered, or it represents the trajectory of a new pole. The latter possibility would spoil Reggeisation, for, to provide the missing zero order term in the expansion of this new Regge pole, we should have to re-introduce a Born approximation term, giving non-analytic behaviour. Thus it is essential for Reggeisation that the first alternative should hold. The matter can only be decided by looking at the structure of higher order terms, such as those corresponding to the diagrams of Fig. 3.9.5. An outline discussion confirms that the behaviour of these diagrams corresponds to the alternative which preserves Reggeisation.

If a similar analysis is applied to a *scalar* particle interacting with a neutral vector meson through a conserved current (Gell-Mann, Goldberger, Low, Singh & Zachariasen, 1964) it is found that the scalar particle does not lie on the trajectory associated with $l = 0$.

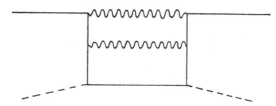

Fig. 3.9.4. Another diagram related to Reggeization.

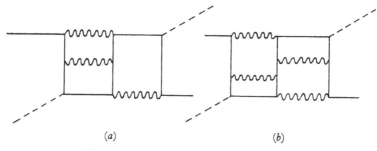

(a)   (b)

Fig. 3.9.5. Diagrams related to Fig. 3.9.4.

Thus Reggeisation shows a remarkable difference of behaviour between particles of spin 0 and particles of spin $\frac{1}{2}$. It also provides a counter-example to the otherwise plausible conjecture that particles which have fundamental fields associated with them must correspond to non-analytic behaviour in complex angular momentum.

3.10 Production processes

Introduction

So far we have considered two-particle scattering without the production of further particles. However, both for phenomenological purposes, and also for the determination of subtractions in possible eventual dispersion representations, it is desirable to know about the high energy behaviour of production processes. These amplitudes are even more difficult than scattering amplitudes to discuss by rigorous methods, but the methods developed for the study of perturbation theory models are readily applicable (Halliday & Polkinghorne, 1963; Polkinghorne, 1965a).

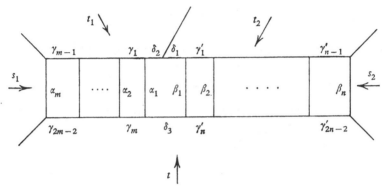

Fig. 3.10.1. A two-fold Regge pole.

It is clearly easy to obtain single Regge pole and cut contributions of the forms discussed in the preceding sections simply by attaching appropriate external lines to the ladder and other diagrams considered. In this section therefore we shall restrict our explicit discussion to diagrams which are characteristic of production processes and which do not have analogues in the case of scattering amplitudes. A typical example is provided by Fig. 3.10.1 which illustrates a *two-fold* pole occurring in a two-particle to three-particle scattering amplitude.

Two simple limits

The behaviour associated with Fig. 3.10.1 will depend upon which limit is being evaluated; that is to say, how the subenergies t_1 and t_2 behave as the total energy $t \to \infty$. We shall first consider two special cases by means of the approximation technique and then give a discussion of a third case by the use of an appropriate generalisation of the Mellin transform method.

(i) $t \to \infty$; t_1, t_2, s_1, s_2, all fixed. The contribution of Fig. 3.10.1 is of the form

$$g^5 \pi^2 r! \left(\frac{-g^2}{16\pi^2} \right)^{r-2} \int_0^1 \Pi d\xi \frac{[C(\xi)]^{r-1} \delta(\Sigma\xi - 1)}{[D(t, t_1, t_2, s_1, s_2; \xi)]^{r+1}}, \qquad (3.10.1)$$

where $r = m + n$, and ξ is a collective symbol for the Feynman parameters of the figure.

The coefficient of t in D is simply

$$\alpha_1 \ldots \alpha_m \beta_1 \ldots \beta_n, \tag{3.10.2}$$

and so the leading behaviour is associated with the neighbourhood of $\alpha_1 = \ldots = \beta_n = 0$. A straightforward leading approximation calculation then gives

$$g^2 [K(s_1)]^{m-1} [K(s_2)]^{n-1} \beta(s_1, s_2) \, t^{-1} \frac{(\log t)^{r-1}}{(r-1)!}, \tag{3.10.3}$$

with $K(s)$, the ladder function, defined by (3.3.9), and

$$\beta(s_1, s_2) = \pi^2 g^3 \int_0^1 \frac{d\delta_1 d\delta_2 d\delta_3 \, \delta(\delta_1 + \delta_2 + \delta_3 - 1)}{[s_1 \delta_2 \delta_3 + s_2 \delta_3 \delta_1 + \delta_1 \delta_2 - 1]^2}, \tag{3.10.4}$$

with the masses of the particles all taken equal to unity. Summing (3.10.3) over m and n for fixed r yields

$$g^2 \beta(s_1, s_2) \frac{[K(s_1)]^{r-1} - [K(s_2)]^{r-1}}{K(s_1) - K(s_2)} \, t^{-1} \frac{(\log t)^{r-1}}{r!}, \tag{3.10.5}$$

which in turn summed over r gives as asymptotic behaviour

$$g^2 \beta(s_1, s_2) \frac{t^{\alpha(s_1)} - t^{\alpha(s_2)}}{\alpha(s_1) - \alpha(s_2)}. \tag{3.10.6}$$

As usual, $$\alpha(s) = -1 + K(s). \tag{3.10.7}$$

The behaviour (3.10.6) is just what is obtained as the leading behaviour associated via the Watson–Sommerfeld transform with the two-fold pole

$$\frac{g^2 \beta(s_1, s_2)}{[l - \alpha(s_1)] [l - \alpha(s_2)]}. \tag{3.10.8}$$

(ii) $t_1 = k_1 t$; $t_2 = k_2 t$; $t \to \infty$; s_1, s_2, fixed. The coefficient of the asymptotic variable t is now

$$\alpha_1 \ldots \alpha_m \beta_1 \ldots \beta_n + \alpha_1 \ldots \alpha_m \Delta_1(\beta, \gamma', \delta_1) k_1 + \beta_1 \ldots \beta_n \Delta_2(\alpha, \gamma, \delta_2) k_2, \tag{3.10.9}$$

where
$$\left. \begin{aligned} \Delta_1 &= \delta_1 \Delta_n(\beta, \gamma') + \beta_1 \Delta_n'(\beta, \gamma'), \\ \Delta_2 &= \delta_2 \Delta_m(\alpha, \gamma) + \alpha_1 \Delta_m'(\alpha, \gamma). \end{aligned} \right\} \tag{3.10.10}$$

The functions $\Delta_n \ldots \Delta_m'$ can be written down explicitly but all that is relevant for our analysis is that they have the properties (a) Δ_1 does not vanish when one of the β's is zero, (b) when all the β's are zero

$$\Delta_1 = \delta_1 \prod_{i=1}^{n-1} (\gamma_i' + \gamma_{i+n-1}'); \tag{3.10.11}$$

together with analogous properties for Δ_2.

The coefficient (3.10.9) possesses a number of distinct equivalent scaling sets in the way discussed in § 3.4. They can be enumerated in the following way:

We first scale δ_1, β_1; δ_2, α_1; α_1, β_1. The δ-functions associated with these scalings put $\bar{\delta}_1 = \bar{\delta}_2 = 1$, $\bar{\alpha}_1 + \bar{\beta}_1 = 1$. Subsequent scalings can then be determined by a rule. Suppose that the immediately previous scaling involved $\bar{\alpha}_i$ and $\bar{\beta}_j$. (We start, of course, with $i = j = 1$.) If $i < m$, $j < n$ we *either* scale $\bar{\alpha}_i$ with β_{j+1}, so that linearisation in $\delta(\bar{\alpha}_i + \bar{\beta}_j - 1)$ forces $\bar{\beta}_j = 1$, or we scale $\bar{\beta}_j$ with α_{i+1}, forcing $\bar{\alpha}_i = 1$. If $i = m$, $j < n$ we scale $\bar{\alpha}_i$ with β_{j+1}, forcing $\bar{\beta}_j = 1$. If $i < m$, $j = n$ we scale $\bar{\beta}_j$ with α_{i+1}, forcing $\bar{\alpha}_i = 1$. If $i = m$, $j = n$, no further scaling is possible.

The alternatives given in this rule enumerate and exhaust the independent scalings which are possible and each particular set of choices is found to lead to the same contributions to the asymptotic behaviour. The number of these distinct scalings is clearly the number of different orders in which one sets

$$\bar{\alpha}_1 = \ldots = \bar{\alpha}_{m-1} = 1, \quad \bar{\beta}_1 = \ldots = \bar{\beta}_{n-1} = 1,$$

with the condition that $\bar{\alpha}_i$ must not be put equal to 1 before $\bar{\alpha}_{i'}$ if $i > i'$, and similarly $\bar{\beta}_j$ must not be put equal to 1 before $\bar{\beta}_{j'}$ if $j > j'$. This number is just the binomial coefficient

$$\frac{(n+m-2)!}{(m-1)!\,(n-1)!}. \tag{3.10.12}$$

Thus the leading behaviour is found to be

$$\pi^2 g^5 k_1^{-1} k_2^{-1} [K(s_1)]^{m-1} [K(s_2)]^{n-1} \frac{(m+n-2)!}{(m-1)!\,(n-1)!}\, t^{-2} \frac{(\log t)^{n+m}}{(n+m)!}. \tag{3.10.13}$$

When (3.10.13) is summed over m and n it gives a leading behaviour

$$\frac{\pi^2 g^5}{k_1 k_2} \frac{t^{\alpha(s_1)+\alpha(s_2)}}{[K(s_1)+K(s_2)]^2}. \tag{3.10.14}$$

Physical region limit

Neither of the limits (i) and (ii) is possible in the physical region of the production process although, of course, they would be relevant to the question of subtractions in a possible spectral theory. Ivanter, Popova & Ter-Martirosyan (1964) showed that in the physical region for fixed s_1 and s_2 the subenergies t_1 and t_2 must tend to infinity in such a way that their product is of the order of t. We thus consider the third limit

(iii) $t_1 \sim t^{x_1}$; $t_2 \sim t^{x_2}$; $x_1 + x_2 = 1$; $t \to \infty$; s_1, s_2 fixed. The method of approximation used for (i) and (ii) is not easily adaptable to this case where fractional exponents occur. Instead we use Mellin transforms (Polkinghorne, 1965a). If the transform of (3.10.1) is taken with respect to $-t$, $-t_1$, $-t_2$, with transform variables a, a_1, a_2 respectively, we obtain

$$L(a, a_1, a_2) = \pi^2 g^5 \left(\frac{-g^2}{16\pi^2}\right)^{r-2} (-1)^{r+1} \Gamma(-a) \Gamma(-a_1) \Gamma(-a_2)$$

$$\times \int_0^\infty d\xi [\alpha_1 \dots \alpha_m]^{a+a_1} . [\beta_1 \dots \beta_n]^{a+a_2} . \Delta_1^{a_1} \Delta_2^{a_2} C^{-2-a-a_1-a_2} \exp(-J),$$

$$(3.10.15)$$

with $-JC$ equal to D evaluated with $t = t_1 = t_2 = 0$. The inversion formula gives

$$f = \frac{1}{(2\pi i)^3} \int_C da \int_{C_1} da_1 \int_{C_2} da_2 L(a, a_1, a_2) (-t)^a (-t_1)^{a_1} (-t_2)^{a_2}.$$

$$(3.10.16)$$

In (3.10.16) the effective exponent in the limit (iii) is

$$\tilde{a} = a + a_1 x_1 + a_2 x_2, \tag{3.10.17}$$

and we shall obtain the leading behaviour by distorting the contours C, C_1, C_2 so as to obtain the minimum value of $\tilde{a}$. The extent to which this can be done will, of course, be limited by the singularities of L. These singularities arise in the usual way from the Γ functions in (3.10.15) and divergences of the integral at its lower limits of integration. In particular the α_i integration diverges at $\alpha_i = 0$ if

$$a + a_1 = -1$$

and as in §3.6 integration by parts can be used to exhibit the consequent m-fold pole in L at $a + a_1 = -1$. There is a similar n-fold pole at $a + a_2 = -1$ due to the β_j integrations.

The locations of these poles are given by

$$\left.\begin{array}{l} 0 = a + a_1 + 1 = \tilde{a} + (1 - x_1)(a_1 - a_2) + 1, \\ 0 = a + a_2 + 1 = \tilde{a} - x_1(a_1 - a_2) + 1. \end{array}\right\} \tag{3.10.18}$$

If (3.10.15) is regarded as a function of a, a_1, a_2 the rightmost singularity in the left-half a-plane will occur at $a = -1$ due to the pinch of the multiple poles (3.10.18) in the $(a_1 - a_2)$ integration. In order to evaluate the nature of this singularity it is sufficient to evaluate the residue at one of the multiple poles. This is because the pinched integral clearly differs from an integral which has been drawn back across one of the multiple poles (and so is not pinched between them) by just this residue. The resulting expression is found to be

$$g^2\beta(t_1, t_2)\,[x_1 K(s_1)]^{m-1}\,[(1-x_1)\,K(s_2)]^{n-1}\,\frac{(n+m-2)!}{(m-1)!\,(n-1)!}\,(\tilde{a}+1)^{-n+m+1},$$

(3.10.19)

where

$$\beta(t_1, t_2) = \pi^2 g^3 \int_B db\,\Gamma(1+b)\,\Gamma^2(-b)$$

$$\times \int_0^\infty d\delta_1 d\delta_2 d\delta_3(\delta_1 + \delta_2 + \delta_3)^{-1-b}\,\delta_1^b\,\delta_2^b \exp(-j). \quad (3.10.20)$$

The function j is

$$-(\delta_2\delta_3 s_1 + \delta_3\delta_1 s_2 + \delta_1\delta_2)\,(\delta_1+\delta_2+\delta_3)^{-1} + \delta_1 + \delta_2 + \delta_3. \quad (3.10.21)$$

The variable b is just $\frac{1}{2}(a_1 + a_2)$ and its contour B runs parallel to the imaginary axis in the left half plane.

The asymptotic behaviour obtained by summing (3.10.19) over m and n is

$$g^2 B(t_1, t_2)\,.\,(-t)^{x_1\alpha(s_1)+x_2\alpha(s_2)}. \quad (3.10.22)$$

Such a form was proposed heuristically by Kibble (1963) and Ter-Martirosyan (1963).

The Mellin transform method can also be used to obtain the asymptotic behaviour in a wide variety of limits (Polkinghorne, 1965 a).

CHAPTER 4

S-MATRIX THEORY

4.1 Introductory survey

We have given in §1.1 reasons for wanting to develop a theory of the S-matrix that does not necessarily require an underlying field theory. We have listed there some of the basic principles on which the theory is to be based and this chapter is devoted to a development of the theory on these lines. Apart from the first paper on the subject by Heisenberg (1943), much of the stimulus for the work has been given by Chew, for example in his book (1962). The first full-scale attempts to carry through such a programme were made by Stapp (1962a, b) and Gunson (1965†). We shall follow the development due to Olive (1964) which made essential use of Gunson's ideas.

In the second, third and fourth sections of the chapter we present the assumptions in detail, elucidating their physical basis as far as we can and deriving their immediate consequences. First, in §4.2, we define the S-matrix in terms of asymptotic states, which we require to obey the superposition principle. We find that the S-matrix is unitary so that, as we saw in §1.2, if its elements are interpreted as probability amplitudes, probability is conserved. In this section we also introduce the connectedness structure of the matrix elements, which arises from the fact that, because of the assumed short-range nature of the fundamental forces, particles can interact among themselves in clusters.

§4.3 is mainly devoted to the consequences of Lorentz invariance. We show how the matrix elements may be expressed as functions of Lorentz scalars, rather than of the four-momenta in terms of which they are initially defined. We discuss the various constraints and inequalities placed upon the scalars by the requirement that the momenta be physical.

As we noted in §1.1, an important physical ingredient of the theory is believed to be that of causality. This is usually supposed to result in analyticity properties, but exactly how is not at present clear. So in §4.4 we present the property of analyticity as a postulate. We formulate it in as weak a way as possible and try to interpret its consequences

† This paper was available in preprint form in 1963.

in terms of physics wherever we can in later work. The basic idea is that the physical amplitudes are boundary values of analytic functions whose only singularities are those required by unitarity, the simplest such singularities being, as we saw in § 1.3, the normal thresholds. An important part of the analyticity postulate is the '$i\epsilon$-prescription', which tells us from which side we must approach the cuts attached to the normal-threshold singularities in order to achieve the physical values of the matrix elements; we examine this in some detail and conclude that we must adopt the same prescription as operates in perturbation theory.

§ 4.5 deals with the physical-region singularities corresponding to single stable particles. The existence of these singularities in the multiparticle amplitudes may, similarly to the case of normal thresholds, be deduced directly from unitarity. Their nature is closely connected with causality and we show that, as in perturbation theory, they are poles whose residues are the products of lower amplitudes.

In § 4.6 we show how the unitarity equations give formulae for the discontinuities associated with the physical-region normal-threshold singularities. To obtain the discontinuities associated with branch-points outside (or on the edge of) the physical region, we need to have formulae analogous to the unitarity equations but operating outside the physical region. How such equations may be proved is indicated by detailed discussion of a particular example, the hermitian analyticity equation for the two-particle → two-particle amplitude. (The importance of this equation has already been explained in § 1.3). The equation is proved by embedding the amplitude under study in a higher amplitude, through its occurrence as a factor in the residue of a single-particle pole in that amplitude, and analysing the physical unitarity equation for the higher amplitude.

§ 4.7 is designed to give greater insight into the structure of the normal-threshold singularities occurring in the several variables associated with a multiparticle amplitude. We show that the two-particle normal thresholds are square-root in nature, and give, without proof, some discontinuity formulae and examine their properties. By seeing how to add formulae for various individual discontinuities we verify that the unitarity equations for multiparticle amplitudes give the simultaneous discontinuities associated with several singularities in several variables.

In § 4.8 we deduce that the ingredients introduced into the theory so far require that to every particle there correspond an antiparticle with equal mass and opposite additive quantum numbers. We also prove the crossing property described in § 1.3, and the TCP theorem. This is done again by embedding the amplitude under study in a higher amplitude through its occurrence as a factor in the residue of a pole.

So far mention has been made only of stable particles. In § 4.9 unstable particles are introduced and it is shown that, mathematically, they enter the theory in a very similar way. Their properties are most simply discovered by means of a model, which is discussed in some detail. An example is given of how the predictions of the model may be verified rigorously.

The work of §§ 4.6–4.9 explicitly considers only the normal thresholds. In § 4.10 we show that unitarity generates further singularities and leads to an analytic structure very similar to that of finite-order perturbation theory. Thus the work of chapter 2 now provides important insight. In § 4.11 a particular example, a triangle singularity occurring in the physical region of the three-particle → three-particle amplitude, is shown from unitarity alone to display the same physical-region properties that were discovered in the perturbation-theory analysis of § 2.3.

Once the presence of the extra singularities is established, it is necessary to verify that they do not interfere with the results earlier deduced, such as crossing and hermitian analyticity. This cannot be done without more knowledge of the structure of the singularities, so here (in § 4.11) we only consider the effect of the triangle singularity on the derivation of hermitian analyticity.† More work on this will presumably be done in the future; we are optimistic about its results because of the similarity of the singularity structure with that of perturbation theory in which, as far as perturbation theory has a meaning, the theorems are believed to be true. An encouraging feature of perturbation-theory singularities, which we made use of in the general discussion of § 2.6, is that successively more complicated singularities may be discussed one by one. The status of this 'hier-archical principle' in S-matrix theory is not clear except, as we indicate in § 4.11, in the physical region, but one might hope that it is possible to justify the procedure of starting with the normal thresholds and then generating successively more complicated singu-larities.

---

† Its effect on the proof of crossing has been considered by Stapp (1965).

4.2 Unitarity and connectedness-structure

The asymptotic states and the S-matrix

In a scattering experiment, one first observes particles moving towards each other and later one observes them moving apart. If these observations are made, respectively, sufficiently early and sufficiently late, the particles will be far apart. So if the forces between the particles are suitably weak at large distances, the particles behave essentially as free particles during the observations, apart from the disturbances they suffer as a result of interaction with the measuring apparatus. We are therefore able to make use of the familiar quantum-mechanical description of free particles in terms of state vectors $|\rangle$ (Dirac, 1958).

Accordingly we assume that in the extreme past $(t \to -\infty)$ the particles may be described by states $|\text{in}\rangle$ with the following properties:

(i) *The superposition principle.* If $|\psi, \text{in}\rangle$ and $|\phi, \text{in}\rangle$ are physically occurring states, so is $\lambda|\psi, \text{in}\rangle + \mu|\phi, \text{in}\rangle$, where λ and μ are any complex numbers. (We shall not discuss the modifications resulting from the operation of a superselection rule.)

(ii) The set of physical states is normalizable and complete, so that a subset may be chosen that satisfies the *orthonormality* conditions

$$\langle m, \text{in}|\, n, \text{in}\rangle = \delta_{mn}, \tag{4.2.1a}$$

and the *completeness* relation

$$\sum_m |m, \text{in}\rangle\langle m, \text{in}| = 1, \tag{4.2.1b}$$

where m and n are labels describing particular configurations of free particles.

Similarly, we assume that in the infinite future $(t \to +\infty)$ the particles may be described by states $|\text{out}\rangle$, bearing corresponding labels and with exactly similar properties.

We define an operator S by

$$S = \sum_m |m, \text{in}\rangle\langle m, \text{out}|, \tag{4.2.2a}$$

so that

$$S^\dagger = \sum_m |m, \text{out}\rangle\langle m, \text{in}|. \tag{4.2.2b}$$

Because of the orthonormality and completeness conditions (4.2.1) on the states $|m, \text{in}\rangle$, and the analogous conditions on the states $|m, \text{out}\rangle$, the matrix elements of S satisfy

$$\langle \phi, \text{in}|\, S|\psi, \text{in}\rangle = \langle \phi, \text{out}|\,\psi, \text{in}\rangle = \langle \phi, \text{out}|\, S|\psi, \text{out}\rangle, \tag{4.2.3}$$

where the labels ϕ, ψ refer to any physical state. Further, the ortho-normality and completeness conditions imply that S is unitary:

$$SS^\dagger = 1, \qquad\qquad (4.2.4a)$$

and
$$S^\dagger S = 1. \qquad\qquad (4.2.4b)$$

We saw in §1.2 that, if we make the assumption that the matrix element (4.2.3) is a probability amplitude for transition from the state labelled by ψ to that labelled by ϕ, that is if the square of the modulus of the matrix element is the probability for the occurrence of the transition, the unitarity condition corresponds to the physical requirement of probability conservation.

Momentum eigenstates

A complete set of physical states that is of particular interest is the set of momentum eigenstates. It was pointed out by Heisenberg (1943), in his original paper on the S-matrix, that the energy-momentum of a particle may, in principle, be measured to arbitrary accuracy. This is because a measurement of the time of flight T between two detectors a distance X apart yields for the velocity of a particle the expression

$$\frac{X + \mathscr{E}X}{T + \mathscr{E}T}.$$

Here $\mathscr{E}X$ and $\mathscr{E}T$ are the errors in the measurement of X and T arising from the uncertainty principle, which does not allow precise measurement of these quantities if the momentum is not to be altered by the measurement. The errors $\mathscr{E}X$ and $\mathscr{E}T$ need not depend on X and T, so if we make X and T very large the errors become negligible by comparison and we obtain the velocity to arbitrary accuracy. If the mass of the particle is precisely known (which is open to some doubt; see Eden & Landshoff, 1965), knowledge of the velocity implies knowledge of the energy-momentum.

Because the momentum eigenstates are accessible to experiment, and because the corresponding S-matrix elements have particularly interesting properties, we concentrate our attention on these states. For simplicity we consider the theory of a single type of spin-zero particle obeying Bose statistics.†

Using the standard annihilation-creation operator formalism, we

† *Note added in proof.* The connection between spin and statistics has been derived by Lu & Olive (1966) using the general-spin analysis of J. R. Taylor (1966).

write the single particle state $|p\rangle$ describing a particle with momentum $\mathbf{p}$ and energy $p^0 = +\sqrt{(\mathbf{p}^2 + \mu^2)}$, as

$$|p\rangle = a^\dagger(\mathbf{p})|0\rangle,$$

where $|0\rangle$ is the vacuum state with no particles, i.e.

$$a(\mathbf{p})|0\rangle = 0, \quad \text{all } \mathbf{p}.$$

We shall choose the relativistic normalisation

$$\langle p'|p\rangle = [a(\mathbf{p}'), a^\dagger(\mathbf{p})] = \Delta(\mathbf{p}', \mathbf{p}), \tag{4.2.5}$$

where $\qquad \Delta(\mathbf{p}', \mathbf{p}) = (2\pi)^3 \, 2p_0 \, \delta^{(3)}(\mathbf{p}' - \mathbf{p}). \tag{4.2.6a}$

We note that there is an arbitrary phase in $a(\mathbf{p})$ since its properties are invariant under the phase transformation $a(\mathbf{p}) \to e^{i\phi} a(\mathbf{p})$, with ϕ independent of $\mathbf{p}$.

When previously we had discrete states with unit normalisation we used the summation property

$$\sum_b f_b \delta_{ab} = f_a.$$

The integration yielding the corresponding property with $\Delta(\mathbf{p}', \mathbf{p})$,

$$\underset{\mathbf{p}}{S} f(\mathbf{p}) \Delta(\mathbf{p}', \mathbf{p}) = f(\mathbf{p}'),$$

is given by $\qquad \underset{\mathbf{p}}{S} = \dfrac{1}{(2\pi)^3} \displaystyle\int \dfrac{d^3\mathbf{p}}{2p_0} = (2\pi)^{-3} \int d^4p\, \delta^{(+)}(p^2 - \mu^2). \tag{4.2.6b}$

We define the state of n particles with momenta $p_1, p_2, \ldots, p_n$ as

$$|p_1, p_2, \ldots, p_n\rangle = a^\dagger(\mathbf{p}_1)\, a^\dagger(\mathbf{p}_2) \ldots a^\dagger(\mathbf{p}_n)\,|0\rangle. \tag{4.2.7}$$

Because of (4.2.5) the normalisation of this state is

$$\langle p_1', p_2', \ldots, p_n' | p_1, p_2, \ldots, p_n\rangle = \sum_{(\nu)} \{\Delta(\mathbf{p}_1', \mathbf{p}_{\nu_1}) \ldots \Delta(\mathbf{p}_n', \mathbf{p}_{\nu_n})\}, \tag{4.2.8}$$

while states containing different numbers of particles are orthogonal. In (4.2.8) the summation is over the $n!$ ways of pairing off the momenta in the two states. The corresponding n-particle phase integral is

$$\underset{\mathbf{p}_1 \ldots \mathbf{p}_n}{S} = \frac{1}{n!} \underset{\mathbf{p}_1}{S} \ldots \underset{\mathbf{p}_n}{S} \tag{4.2.9}$$

so that the completeness relation reads

$$\sum_{n=0}^{\infty} \frac{1}{(2\pi)^{3n}\, n!} \int d^4q_1 \ldots d^4q_n\, \delta^{(+)}(q_1^2 - \mu^2) \ldots \delta^{(+)}(q_n^2 - \mu^2)$$
$$\times \, |q_1, \ldots, q_n\rangle \langle q_1, \ldots, q_n| = 1. \tag{4.2.10}$$

Bubble notation

We are concerned with matrix elements

$$\langle p'_1, ..., p'_{m'} | S | p_1, ..., p_m \rangle \tag{4.2.11}$$

that describe transitions of a configuration of m particles, each with definite momentum, into one of m' particles. Because of (4.2.3) the state vectors in (4.2.11) may be either of 'in' type or of 'out' type, so henceforth we do not write in these labels.

A basic assumption of the theory is that the total energy-momentum is conserved in the interaction; the matrix-element vanishes unless

$$\Sigma p = \Sigma p'. \tag{4.2.12}$$

This has an important simplifying effect when the completeness relation (4.2.10) is used to express the unitarity relations (4.2.4) in terms of the matrix elements of the type (4.2.11). The relations derived from (4.2.4a) read

$$\sum_n \frac{1}{(2\pi)^{3n} n!} \int \prod_{i=1}^{n} d^4 q_i \, \delta^{(+)}(q_i^2 - \mu^2) \langle p'_1, ..., p'_{m'} | S | q_1, ..., q_n \rangle$$

$$\times \langle q_1, ..., q_n | S^\dagger | p_1, ..., p_m \rangle = \langle p'_1, ..., p'_{m'} | p_1, ..., p_m \rangle, \tag{4.2.13}$$

while those derived from (4.2.4b) have the operators S, $S^\dagger$ interchanged. Because of energy-momentum conservation only the states with

$$\Sigma p = \Sigma q = \Sigma p' \tag{4.2.14}$$

survive in the integration-summation. In particular, the summation is restricted to values of n that satisfy

$$(n\mu)^2 \leqslant (\Sigma p)^2. \tag{4.2.15}$$

It is convenient to introduce a diagrammatic notation for the equations (4.2.13). We write, for the matrix elements of S,

$$\langle p'_1, ..., p'_{m'} | S | p_1, ..., p_m \rangle \equiv \quad \tag{4.2.16}$$

and similarly for the matrix-elements of $S^\dagger$. Also we write

$$\langle p'_1, ..., p'_m | p_1, ..., p_m \rangle = \quad \tag{4.2.17}$$

where there are m lines (which, according to (4.2.9), denote a sum of $m!$ terms in which the momenta are paired off in different ways). Also we write

$$\frac{1}{(2\pi)^{3n}\, n!} \int \prod_{i=1}^{n} (d^4 q_i\, \delta^{(+)}(q_i^2 - \mu^2)) = \;\;\;\;\;\;\;\;\;\;\;\;\;\; . \tag{4.2.18}$$

There will be no confusion between (4.2.17) and (4.2.18), since the lines in the latter will always join two bubbles.

In this notation the unitarity equations (4.2.13) read, when account is taken of (4.2.15), as follows:

$(2\mu)^2 \leqslant (\Sigma p)^2 < (3\mu)^2$:

$$\text{[S]} - \text{[S']} \;=\; = \;; \tag{4.2.19a}$$

$(3\mu)^2 \leqslant (\Sigma p)^2 < (4\mu)^2$:

$$\left.\begin{aligned}
\text{[S]} - \text{[S']} + \text{[S]} = \text{[S']} &= =\,, \\
\text{[S]} - \text{[S']} + \text{[S]} = \text{[S']} &= 0\,, \\
\text{[S]} - \text{[S']} + \text{[S]} = \text{[S']} &= 0\,, \\
\text{[S]} - \text{[S']} + \text{[S]} = \text{[S']} &= \equiv\,;
\end{aligned}\right\} \tag{4.2.19b}$$

$(4\mu)^2 \leqslant (\Sigma p)^2 < (5\mu)^2$:

$$\left.\begin{aligned}
\text{[S]} - \text{[S']} + \text{[S]} = \text{[S']} + \text{[S]} = \text{[S']} &= =\,, \\
\text{[S]} - \text{[S']} + \text{[S]} = \text{[S']} + \text{[S]} = \text{[S']} &= 0\,, \\
\text{[S]} = \text{[S']} + \text{[S]} = \text{[S']} + \text{[S]} = \text{[S']} &= \equiv\,, \\
\text{[S]} = \text{[S']} + \text{[S]} = \text{[S']} + \text{[S]} = \text{[S']} &= \equiv\,, \\
\text{[S]} - \text{[S']} + \text{[S]} = \text{[S']} + \text{[S]} = \text{[S']} &= 0\,, \\
\text{[S]} = \text{[S']} + \text{[S]} = \text{[S']} + \text{[S]} = \text{[S']} &= 0\,, \\
\text{[S]} = \text{[S']} + \text{[S]} = \text{[S']} + \text{[S]} = \text{[S']} &= 0\,, \\
\text{[S]} = \text{[S']} + \text{[S]} = \text{[S']} + \text{[S]} = \text{[S']} &= 0\,, \\
\text{[S]} = \text{[S']} + \text{[S]} = \text{[S']} + \text{[S]} = \text{[S']} &= \equiv\,;
\end{aligned}\right\} \tag{4.2.19c}$$

and so on for larger values of $(\Sigma p)^2$. The equations corresponding to the other form (4.2.4b) of the unitarity condition are obtained by interchanging the labels S, $S^\dagger$ in (4.2.19).

Notice that nowhere have bubbles of the form ⊐○- appeared. This is because the particles we have so far introduced into the theory are supposed to be stable (otherwise they would not exist in the infinite future and so would not appear in the asymptotic states), so that the process that would be represented by such a bubble is not permitted by energy-momentum conservation.

Connectedness structure

An important assumption that must now be introduced into the theory is that of the connectedness-structure, or cluster-decomposition, of the matrix elements. Although it is believed to be a secure assumption, we must emphasize that its precise basis is not fully understood.

We have already displayed part of the connectedness-structure in (1.2.13). For two-particle → two-particle scattering this equation takes the form

$$\text{⊃}(s)\text{⊏} \;=\; \text{⸗} \;+\; \text{⊃}(+)\text{⊏}. \tag{4.2.20a}$$

This structure arises from the short-range nature of the forces; the term ⸗ corresponds to the particles 'missing' each other. For a three-particle → three-particle interaction the particles may all miss one another, or it may be that two of the particles interact, while the third misses them. This leads to the structure

$$\text{⊃}(s)\text{⊏} \;=\; \text{≡} \;+\; \Sigma\,\text{≡⊕≡} \;+\; \text{⊃}(+)\text{⊏}. \tag{4.2.20b}$$

For a four-particle → four-particle interaction there is the further possibility of the particles interacting in pairs, and we have the structure

$$\text{⊃}(s)\text{⊏} \;=\; \text{≣} \;+\; \Sigma\,\text{≣⊕≣} \;+\; \Sigma\,\text{≣⊕≣} \;+\; \Sigma\,\text{≣⊕≣} \;+\; \text{⊃}(+)\text{⊏}. \tag{4.2.20c}$$

In these equations the summations are over the different possible choices of particles that collide and miss. Matrix elements for different numbers of particles in initial and final states can be similarly decomposed, though for these there is no part corresponding to complete absence of interaction:

$$
\left.
\begin{aligned}
\text{⊃}(s)\text{⊏} &= \text{⊃}(+)\text{⊏}, \\
\text{⊃}(s)\text{⊐} &= \text{⊃}(+)\text{⊐}, \\
\text{⊃}(s)\text{⊏} &= \text{⊃}(+)\text{⊏}, \\
\text{⊐}(s)\text{⊐} &= \text{⊐}(+)\text{⊐}, \\
\text{⊐}(s)\text{⊏} &= \Sigma\,\text{≡⊕≡} + \text{⊃}(+)\text{⊏}, \\
\text{⊐}(s)\text{⊏} &= \Sigma\,\text{≡⊕≡} + \text{⊃}(+)\text{⊏}.
\end{aligned}
\right\} \tag{4.2.20d}
$$

In the equations (4.2.20) we have introduced the 'connected parts' of the matrix-elements of S:

$$\langle p'_1, ..., p'_{m'} | S | p_1, ..., p_m \rangle_c = \;\; \genfrac{}{}{0pt}{}{1'}{m'} \!\!\!\!\rightleftharpoons\!\!\!\!\bigoplus\!\!\!\!\rightleftharpoons\!\! \genfrac{}{}{0pt}{}{1}{m}. \qquad (4.2.21\,a)$$

These amplitudes are supposed to be free of the δ-functions explicitly displayed in (4.2.20). We define corresponding connected parts of the matrix-elements of $S^\dagger$

$$\langle p'_1, ..., p'_{m'} | S^\dagger | p_1, ..., p_m \rangle_c = - \;\; \genfrac{}{}{0pt}{}{1'}{m'} \!\!\!\!\rightleftharpoons\!\!\!\!\bigominus\!\!\!\!\rightleftharpoons\!\! \genfrac{}{}{0pt}{}{1}{m}. \qquad (4.2.21\,b)$$

For reasons that appear below, we have chosen to introduce a factor (-1) in this definition. To obtain the connectedness-structure for matrix-elements of $S^\dagger$ we replace the labels S in (4.2.20) by $S^\dagger$, the labels $(+)$ by $(-)$, and include a factor (-1) each time a $(-)$-bubble appears.

We have said that the precise basis of the connectedness-structure is not clear, though it is easy to understand intuitively. Its importance in S-matrix theory was first noted by Heisenberg (1943), who considered the situation in potential theory. Perhaps the easiest way to understand that it should be valid is to consider perturbation theory, where it may be seen from the Feynman rules of § 1.5 that a given Feynman graph is either connected or consists of two or more disconnected parts. An analysis independent of both potential theory and perturbation theory has been given by Wichmann & Crichton (1963), who study a wave-packet formulation of scattering theory and formulate a cluster-decomposition postulate by considering two scattering experiments separated in space-time. In the limit of the separation being infinite the two experiments should not interfere with each other, and so, they assume, the overall S-matrix element that describes them should factorise into the product of the matrix-elements for the individual experiments. This is obviously an important requirement since, in practice, we are not able to empty the universe in order to conduct an experiment in complete isolation.

The extent to which the connectedness-structure is not understood is that it is not known just how wide is its validity. It is thought to break down, at least in the form we have given, for a Coulomb potential $1/r$. This vanishes at $r = \infty$ much less rapidly that a Yukawa potential $e^{-\mu r}/r$, for which the connectedness-structure is thought to be valid. The reason for the breakdown is that in a theory that contains zero-mass particles, such as photons, the particles can never be separated sufficiently to be effectively free of each other's interaction. We have said in §1.1 that we do not even know how to define the S-matrix elements for such a theory.

The unitarity equations

If we insert into the unitarity equation (4.2.19a) for the two-particle → two-particle S-matrix element the connectedness-structure (4.2.20a) for that element, we obtain

$$\left(= + \,\ominus\!\!\oplus\!\!\ominus\right)\left(= - \,\ominus\!\!\ominus\!\!\ominus\right) \;=\; = \,. \tag{4.2.22}$$

Multiplying out and collecting together terms of like connectedness-structure, we have

$$\left(= - =\right) + \left(\ominus\!\oplus\!\ominus - \ominus\!\ominus\!\ominus - \ominus\!\oplus\!\ominus\!\ominus\right) \;=\; 0\,.$$

The first bracket vanishes trivially, leaving the relation

$$\ominus\!\oplus\!\ominus - \ominus\!\ominus\!\ominus = \ominus\!\oplus\!\ominus\!\ominus\,. \tag{4.2.23}$$

Similarly, consider the last equation in (4.2.19b). If we insert the connectedness-structure of (4.2.20) we obtain

$$\ominus\!\oplus\!\ominus\!\ominus + \left(\equiv + \Sigma\,\overline{\ominus} + \ominus\!\oplus\!\ominus\right)\left(\equiv - \Sigma\,\overline{\ominus} - \ominus\!\ominus\!\ominus\right) \;=\; \equiv\,.$$

The product $\left(\Sigma\,\overline{\ominus}\right)\left(\Sigma\,\overline{\ominus}\right)$ includes terms $\left(\overline{\ominus}\right)\left(\overline{\ominus}\right)$ and $\left(\overline{\ominus}\right)\left(\overline{\ominus}\right)$ which we shall distinguish since the first is disconnected and the second connected. Thus we write

$$\left(\Sigma\,\overline{\ominus}\right)\left(\Sigma\,\overline{\ominus}\right) \;=\; \Sigma\,\overline{\ominus\!\ominus} \;+\; \Sigma\,\overline{\ominus\!\ominus}\,.$$

On multiplying out and rearranging to collect together terms of like connectedness-structure we find

$$\left(\equiv - \equiv\right) + \Sigma\left(\overline{\ominus} - \overline{\ominus} - \overline{\ominus\!\ominus}\right)$$

$$+ \left(\ominus\!\oplus\!\ominus - \ominus\!\ominus\!\ominus - \ominus\!\oplus\!\ominus\!\ominus - \ominus\!\oplus\!\ominus\!\ominus - \Sigma\,\overline{\ominus\!\ominus}\!\ominus - \Sigma\,\ominus\!\oplus\!\ominus\!\overline{\ominus}\right.$$

$$\left. - \Sigma\,\overline{\ominus\!\oplus\!\ominus}\right) \;=\; 0\,. \tag{4.2.24}$$

The expressions in each bracket must vanish individually, being of different connectedness-structure and so containing different sets of δ-functions. The first again vanishes trivially; to see that the second vanishes we divide out the factor represented by the upper line and use (4.2.23). The vanishing of the third bracket gives a new equation.

It is not by accident that the second bracket vanishes automatically. It happens because of the specific form of our connectedness-structure assumptions. If we had introduced different phases in (4.2.20), for example

$$\{S\} = \equiv + \Sigma\, e^{i\alpha}\, \overline{\oplus} + \{+\},$$

so that

$$\{S'\} = \equiv - \Sigma\, e^{-i\alpha}\, \overline{\oplus} - \{-\},$$

we should have obtained for the second bracket in (4.2.24)

$$\left(e^{i\alpha}\, \overline{\oplus} - e^{-i\alpha}\, \overline{\ominus} - \overline{\oplus\ominus} \right).$$

In order that the vanishing of this be consistent with (4.2.23), we require that $e^{i\alpha} = 1$ (Stapp, 1962b).

We may treat in a similar manner all the equations (4.2.19), and obtain the following set of equations:

$(2\mu)^2 \leqslant (\Sigma p)^2 < (3\mu)^2$:

$$\{+\} - \{-\} = \{+\}\{-\}\,;\tag{4.2.25a}$$

$(3\mu)^2 \leqslant (\Sigma p)^2 < (4\mu)^2$:

$$\{+\} - \{-\} = \{+\}\{-\} + \{+{=}-\},$$

$$\{+\} - \{-\} = \{+\}\{-\} + \{+{=}-\} + \Sigma\{+{=}\circ\},$$

$$\{+\} - \{-\} = \{+\}\{-\} + \{+{=}-\} + \Sigma\,\overline{\oplus}\{-\},$$

$$\{+\} - \{-\} = \{+\}\{-\} + \{+{=}-\} + \Sigma\,\overline{\oplus}\{-\} + \Sigma\{+\}{=}\circ + \Sigma\,\overline{\oplus{=}\circ}\,;\tag{4.2.25b}$$

$(4\mu)^2 \leqslant (\Sigma p)^2 < (5\mu)^2$:

$$\{+\} - \{-\} = \{+\}\{-\} + \{+{=}-\} + \{+{=}-\},$$

$$\{+\} - \{-\} = \{+\}\{-\} + \{+{=}-\} + \Sigma\{+{=}\circ\} + \{+{=}-\} + \Sigma\{+{=}\circ\},$$

$$\{+\} - \{-\} = \{+\}\{-\} + \{+{=}-\} + \Sigma\,\overline{\oplus}\{-\} + \{+{=}-\} + \Sigma\,\overline{\oplus}\{-\},$$

$$\{+\} - \{-\} = \{+\}\{-\} + \{+{=}-\} + \Sigma\,\overline{\oplus}\{-\} + \Sigma\{+\}{=}\circ + \Sigma\,\overline{\oplus{=}\circ}$$
$$+ \{+{=}-\} + \Sigma\,\overline{\oplus}\{-\} + \Sigma\{+\}{=}\circ + \Sigma\,\overline{\oplus{=}\circ},$$

$$\{+\} - \{-\} = \{+\}\{-\} + \{+{=}-\} + \Sigma\{+{=}\circ\} + \{+{=}-\} + \Sigma\{+\}{=}\circ$$
$$+ \Sigma\{+{=}\circ\} - \Sigma\{+{=}\circ\},$$

$$(4.2.25c)$$

and so on for higher values of $(\Sigma p)^2$.

These equations are the fundamental equations of our theory. They have been grouped according to the different energy ranges to emphasis that as $(\Sigma p)^2$ is increased the equations become both more numerous and more complicated.

Notice that the terms on the right-hand side are preceded by a minus sign if the number of bubbles labelled by $(-)$ is even. Similar equatians, derived from $(4.2.4b)$ instead of $(4.2.4a)$, hold with the $(+)$ and $(-)$ labels interchanged on the right-hand side (though not on the left), provided sign adjustments are made to conform with the above rule.

A-matrix elements

We have not yet extracted all the δ-functions from the matrix elements. It is a familiar consequence of the Feynman rules in perturbation theory (§1.5) that over-all energy-momentum con-servation $(4.2.12)$ leads to a factor $\delta^{(4)}(\Sigma p' - \Sigma p)$ in the connected part $(4.2.21)$ of the S-matrix element. Since energy-momentum con-servation is believed to result from the translational invariance of the theory in space-time, the presence of this $\delta^{(4)}$-function outside the frame-work of perturbation theory could presumably be deduced from an analysis that introduces space-time into the theory, by consideration of states that correspond to wave-packets rather than to the plane waves (momentum eigenstates) we are using. However, this raises considerable difficulties (Eden & Landshoff, 1965), and here we shall be content to *assume* the presence of the $\delta^{(4)}$-function.

Accordingly we write the matrix-element (4.2.21 a) as

$$\langle S \rangle_c = -i(2\pi)^4 \delta^{(4)}(\Sigma p - \Sigma p')\langle A \rangle, \qquad (4.2.26\,a)$$

and assume that the matrix A is free of all δ-functions. In § 4.3 we shall introduce the further assumption that the matrix-elements of A are actually analytic functions.

By hermitian conjugation of (4.2.26 a) we have

$$\langle S^\dagger \rangle_c = i(2\pi)^4 \delta^{(4)}(\Sigma p - \Sigma p')\langle A^\dagger \rangle. \qquad (4.2.26\,b)$$

The reason for introducing the factor i in (4.2.26) will become clear later when we develop the discontinuity content of the unitarity equations. Given that it is useful to insert this factor, we see now why we introduced the factor (-1) in (4.2.21 b): the $(+)$-bubbles now correspond directly to matrix elements of A, and the $(-)$-bubbles to matrix-elements of $A^\dagger$. Explicitly, this correspondence is

$$m \overline{\underset{-}{\overset{+}{\bigcirc}}} m = -i(2\pi)^4 \delta^{(4)}(\Sigma p - \Sigma p') A_{m'm}^{(\pm)}, \qquad (4.2.27)$$

where

$$A_{m'm}^{(+)} = \langle p_1', ..., p_{m'}' | A | p_1, ..., p_m \rangle, \qquad (4.2.28\,a)$$

$$A_{m'm}^{(-)} = \langle p_1', ..., p_{m'}' | A^\dagger | p_1, ..., p_m \rangle = [A_{mm'}^{(+)}]^*. \qquad (4.2.28\,b)$$

We also recall from (4.2.18) the rules for integration:

$$\left.\begin{array}{l} \text{each internal line} \to (2\pi)^{-3} \displaystyle\int d^4q \, \delta^{(+)}(q^2 - \mu^2), \\[2mm] \text{each } n\text{-particle intermediate state} \to \text{a factor } (n!)^{-1}. \end{array}\right\} \quad (4.2.29)$$

Each term in any of the unitarity equations (4.2.25) contains a factor $-i(2\pi)^4 \delta^{(4)}(\Sigma p - \Sigma p')$, which we may cancel throughout the equation. It is also convenient to redistribute certain other factors. Consider a particular term on the right-hand side of any of the equations (4.2.25). Suppose that this term contains r bubbles and n internal lines. The term then contains n integrations d^4q and r $\delta^{(4)}$-functions. We may integrate out $(r-1)$ of these $\delta^{(4)}$-functions by replacing the q-integrations by integrations over $(n - r + 1) = l$ loop-momenta k, as we did in § 1.5 for Feynman diagrams. The remaining $\delta^{(4)}$-function refers to over-all energy-momentum conservation and is the one we have decided to cancel out, together with its $-i(2\pi)^4$ factor. After this cancellation the powers of i and 2π remaining are

$$(- i)^{r-1} (2\pi)^{4(r-1)} (2\pi)^{-3n} = (- 2\pi i)^n \left[\frac{i}{(2\pi)^4}\right]^l .$$

This makes it convenient to redistribute the factors i and 2π such that $(- 2\pi i)$ is associated with each of the n internal lines and $i/(2\pi)^4$ with each of the l loop integrations. We therefore replace (4.2.27) and (4.2.29) by the new rules (Olive, 1964)

$$m' \overline{\underline{}} \stackrel{+}{\ominus} \overline{\underline{}} m \to A^{(\pm)}_{m'm},$$

each internal line $\to - 2\pi i\, \delta^{(+)}(q^2 - \mu^2),$

each loop $\qquad \to \dfrac{i}{(2\pi)^4} \displaystyle\int d^4k,$ $\qquad\qquad$ (4.2.30)

each n-particle intermediate state joining two bubbles $\to \dfrac{1}{n!}$

4.3   Lorentz invariance and kinematics

The A-matrix element (4.2.28) is a function of the four-momenta of the particles in the initial and final states,

$$A^{(\pm)}_{m'm} = A^{(\pm)}_{m'm}(p'_1, ..., p'_{m'}; p_1, ..., p_m). \qquad (4.3.1)$$

Let Λ be any proper orthochronous Lorentz transformation, so that it satisfies

$$\det \Lambda = 1, \qquad\qquad\qquad (4.3.2a)$$

$$\Lambda^0_0 > 0, \qquad\qquad\qquad (4.3.2b)$$

$$\Lambda^T g \Lambda = g, \qquad\qquad\qquad (4.3.2c)$$

where g is the metric tensor, which we are taking to have the diagonal elements $(1, -1, -1, -1)$. Then our Lorentz-invariance assumption (1.2.8) reads

$$A^{(\pm)}_{m'm}(\Lambda p'_1, ..., p'_{m'}; \Lambda p_1, ..., \Lambda p_m) = A^{(\pm)}_{m'm}(p'_1, ..., p'_{m'}; p_1, ..., p_m). \quad (4.3.3)$$

This has the consequence that $A^{(\pm)}_{m'm}$ is actually a function of the Lorentz invariants that can be formed from the momenta, and in this section we investigate this in some detail. The essential result of the section is (4.3.9) below, which expresses the Lorentz-invariance property in tractable form. The equation expresses $A_{m'm}^{(\pm)}$ in terms of functions of variables similar to the s, t, u introduced in (1.2.10) to describe simple scattering. The treatment here follows that of Olive (1965); see also Asribekov (1962), and Byers & Yang (1964).

Lorentz invariants

We list three types of quantities invariant under the Lorentz transformations of (4.3.2):

    (i) inner products $p_i \cdot p_j$,

    (ii) signs of time components, $\epsilon(p^0) \equiv \mathrm{sign}\,(p^0)$,

    (iii) determinants $\epsilon(p_i, p_j, p_k, p_l) = \epsilon_{\lambda\mu\nu\rho}\, p_i^\lambda p_j^\mu p_k^\nu p_l^\rho$, where $\epsilon_{\lambda\mu\nu\rho}$ is the antisymmetric fourth-order tensor.

Here we are extending the meaning of the symbols p to cover the momenta in both the initial and the final states, so that (i) includes products of the momenta in the different states.

The invariance of (i) follows immediately from (4.3.2c), that of (ii) from (4.3.2b) and the fact that the p are time-like, $p^2 = \mu^2 > 0$. The invariance of (iii) follows from (4.3.2a) and the rules of determinant multiplication; since

$$\epsilon(p, q, r, s) = \begin{vmatrix} p^0 & p^1 & p^2 & p^3 \\ q^0 & q^1 & q^2 & q^3 \\ r^0 & r^1 & r^2 & r^3 \\ s^0 & s^1 & s^2 & s^3 \end{vmatrix} \tag{4.3.4}$$

we have $\qquad \epsilon(\Lambda p, \Lambda q, \Lambda r, \Lambda s) = \det \Lambda \,.\, \epsilon(p, q, r, s).$ (4.3.5)

From (4.3.4) it is easy to show that

$$\epsilon(p_1, p_2, p_3, p_4)\, \epsilon(q_1, q_2, q_3, q_4) = -\det(p_i \cdot q_j), \tag{4.3.6}$$

so that, in particular,

$$\epsilon(p_1, p_2, p_3, p_4) = \pm \sqrt{\{-\det(p_i \cdot p_j)\}}. \tag{4.3.7}$$

Hence knowledge of the quantities of type (i) implies knowledge of those of type (iii), except for an ambiguity of sign. The significance of this ambiguity is that if we make a space reflection $(p^0, \mathbf{p}) \to (p^0, -\mathbf{p})$ of all the momenta, the quantities (i) and (ii) are unchanged, but (iii) change sign. The only reason we need consider (iii) is for the information these quantities give us about the sign. Their absolute numerical values are determined by knowledge of the set (i).

In fact, for any set of momenta, we get sufficient information from just *one* of the quantities (iii). For (4.3.6) tells us that if we know the values of the set (i) and also the sign of any one given $\epsilon(p_1, p_2, p_3, p_4)$, we also know the signs of all other $\epsilon(q_1, q_2, q_3, q_4)$.

Similarly, knowledge of just one member of the set (ii) is sufficient. For, if the momenta are real and on the mass shell,

$$p_i \cdot p_j \gtrless \mu_i \mu_j \qquad (4.3.8)$$

and the $\gtrless$ applies according as the zeroth components of p_i, p_j have like or unlike signs. So if we know the values of the set (i) and the sign of any one of the p^0, we know the signs of all the others. In fact, if the momenta are to be physical, it must be that all the $p^0 > 0$, so that none of the set (ii) appears explicitly in the functions that give the physical matrix-elements.

We show below that the sets (i), (ii) and (iii) are a complete set of invariants, in the sense that knowledge of them implies knowledge of the momenta, to within an over-all proper orthochronous Lorentz transformation. So we see that, for physical momenta, we may write the amplitude (4.3.1) as

$$A(p) = A_s(p_m \cdot p_n) + \epsilon(p_i, p_j, p_k, p_l) A_p(p_m \cdot p_n), \qquad (4.3.9)$$

where p_i, p_j, p_k, p_l are *any* set of four of the momenta and we are still using the symbols p to cover the momenta in both initial and final states. Notice that this decomposition relies on the fact that all $\delta^{(4)}$-functions have been extracted from $A(p)$. For example, $\delta^{(4)}(p)$ cannot be expressed as a function of p^2. The first term in the decomposition, A_s, remains unchanged under space-reflection, and so is the 'parity-conserving' part of A. The other term changes sign under space-reflection and so is absent in strong-interaction theory, where parity is conserved. It is also absent in the two-particle → two-particle amplitude, whether or not parity is conserved in the theory, because energy-momentum conservation allows only three independent momenta. We stress that our discussion assumes that the particles have no spin.

Reconstruction of the momenta and the physical region

We now show how knowledge of the sets (i), (ii) and (iii) of invariants enables us to construct the four-momenta, to within a Lorentz transformation (4.3.2).

Since each vector is on the mass-shell any one, p_1 say, may be transformed by such a Lorentz transformation so as to become $(\epsilon(p_1^0)\sqrt{(p_1{}^2)}, 0, 0, 0)$. Then from knowledge of $(p_1 \cdot p_2)$ we know the zeroth component of p_2, namely $p_2^0 = (p_1 \cdot p_2)/\sqrt{p_1{}^2}$. Hence we know also $|\mathbf{p}_2|$, since $\mathbf{p}_2^2 = (p_2^0)^2 - p_2{}^2$. By a space rotation we can make p_2 take the form $(p_2^0, |\mathbf{p}_2|, 0, 0)$, and then we know p_2 completely. [Such a rotation is a Lorentz transformation of the type (4.3.2) and it does not affect the form we have achieved for p_1.] We may similarly construct p_3, using a rotation to make its fourth component vanish. When we come to construct p_4, its first three components are obtained from $p_4 \cdot p_1$, $p_4 \cdot p_2$ and $p_4 \cdot p_3$. The last component satisfies

$$(p_4^3)^2 = (p_4^0)^2 - (p_4^1)^2 - (p_4^2)^2 - p_4{}^2,$$

which determines it up to a sign, which cannot be varied by a transformation (4.3.2). This sign is fixed by the sign of $\epsilon(p_1, p_2, p_3, p_4)$. Further vectors may be constructed similarly.

In order that all the components of the momenta so constructed be real, there are certain restrictions on the values of the inner products $p_i \cdot p_j$. These restrictions, together with the requirement that the zeroth components of the momenta be positive, determine what is the physical region in the space of the invariants. It is convenient to express these restrictions in terms of Gram determinants, which are defined by

$$G\begin{pmatrix} p_1 \cdots p_N \\ q_1 \cdots q_N \end{pmatrix} = \det(p_i \cdot q_j) \quad (i,j = 1, 2, \ldots, N). \quad (4.3.10)$$

If we use the shorthand notation

$$G\begin{pmatrix} p_i p_j \cdots \\ p_l p_m \cdots \end{pmatrix} = G\begin{pmatrix} ij \cdots \\ lm \cdots \end{pmatrix}, \quad (4.3.11)$$

the results of our construction of the vectors p_1, p_2, p_3, p_4 may be written, after a lengthy calculation, as

$$
\left.
\begin{aligned}
p_1 &= \left(\sqrt{G(\tfrac{1}{1})},\ 0,\ 0,\ 0\right), \\[2mm]
p_2 &= \left(\frac{G(\tfrac{1}{2})}{\sqrt{G(\tfrac{1}{1})}},\ \sqrt{\frac{-G(\tfrac{12}{12})}{G(\tfrac{1}{1})}},\ 0,\ 0\right), \\[2mm]
p_3 &= \left(\frac{G(\tfrac{1}{3})}{\sqrt{G(\tfrac{1}{1})}},\ \frac{-G(\tfrac{12}{13})}{\sqrt{-G(\tfrac{12}{12})\,G(\tfrac{1}{1})}},\ \sqrt{\frac{-G(\tfrac{123}{123})}{G(\tfrac{12}{12})}},\ 0\right) \\[2mm]
p_4 &= \left(\frac{G(\tfrac{1}{4})}{\sqrt{G(\tfrac{1}{1})}},\ \frac{-G(\tfrac{12}{14})}{\sqrt{-G(\tfrac{12}{12})\,G(\tfrac{1}{1})}},\ \frac{-G(\tfrac{123}{124})}{\sqrt{-G(\tfrac{123}{123})\,G(\tfrac{12}{12})}},\ \sqrt{\frac{-G(\tfrac{1234}{1234})}{G(\tfrac{123}{123})}}\right).
\end{aligned}
\right\} \quad (4.3.12)
$$

Hence the restrictions are

$$G\begin{pmatrix}1\\1\end{pmatrix} > 0, \quad G\begin{pmatrix}12\\12\end{pmatrix} < 0, \quad G\begin{pmatrix}123\\123\end{pmatrix} > 0, \quad G\begin{pmatrix}1234\\1234\end{pmatrix} < 0. \quad (4.3.13)$$

These restrictions are unsymmetrical in the labels 1, 2, 3, 4 because of the order in which we chose to construct the momenta. If we had chosen a different order the labels would have appeared differently in the restrictions. But since the set (4.3.13) is evidently both necessary and sufficient, its validity implies also that of the other possible sets, since each vector is on the mass-shell. The truth of this may also be seen by using the Jacobi identity (see, for example, Aitken, 1954, p. 97):

$$D_j^i D_l^k - D_l^i D_j^k = D_{jl}^{ik} D. \quad (4.3.14)$$

Here D_j^i denotes the (i,j) algebraic minor of any determinant D, and D_{jl}^{ik} is the algebraic minor (ik,jl).

When we construct further vectors a new type of restriction enters, that arising from the dimensionality of space. Any five or more four-vectors must be linearly related. (However, because of energy-momentum conservation, the linear relation is trivial unless the sum of the numbers of particles in initial and final states is at least six). By using the scalar product of the linear relation in turn with each of the vectors involved, we obtained a set of homogenous simultaneous linear equations for the coefficients in the relation. The determinants of these equations must vanish, and this determinant is a Gram determinant. So we obtain the conditions

$$\left.\begin{aligned} G\begin{pmatrix}12345\\12345\end{pmatrix} &= 0; \\[1ex] G\begin{pmatrix}12346\\12346\end{pmatrix} &= G\begin{pmatrix}123456\\123456\end{pmatrix} = 0; \\[1ex] G\begin{pmatrix}12347\\12347\end{pmatrix} &= G\begin{pmatrix}123457\\123457\end{pmatrix} = G\begin{pmatrix}1234567\\1234567\end{pmatrix} = 0; \end{aligned}\right\} \quad (4.3.15)$$

and so on as more vectors are constructed. Another way to obtain these relations is to construct the vectors with the dimensionality condition relaxed, and then require that their superfluous components should vanish. This method shows the sufficiency of the conditions (4.3.15).

Choice of variables

Our construction of the vectors p_1, p_2, p_3, p_4 in (4.3.12) results in ten non-zero components. If we construct further vectors they will not, in general, have any zero components. Hence, for $N \geqslant 4$, N four-vectors involve $(4N-6)$ non-zero components. So, taking into account the N on-mass-shell conditions $p^2 = \mu^2$ and the four energy-momentum conservation conditions (4.2.12), we see that an m-particle$\rightarrow m'$-particle process involves $[3(m+m')-10]$ independent variables.

This number is rather less than the number of inner products in the set (i), for two reasons. First, the energy-momentum conservation law (4.2.12) leads to a number of linear relations among the inner products, of which a typical example is that connecting s, t, u in (1.2.12). Secondly, the dimensionality constraints of the type (4.3.15) lead to quadratic relations among the inner products. The fact that the latter relations are quadratic means that, if we select $[3(m+m')-10]$ of the inner products for use as the independent variables, when we solve the relations to obtain the other inner products in terms of our independent set we find ambiguities of sign. Different choices of these signs correspond to different configurations of the momenta, not related by Lorentz transformations. To avoid this difficulty it is best not to make any choice of independent variables, but to think of the amplitudes as functions of all the variables, with the constraints on the variables operating as auxiliary conditions.

In place of the inner products themselves, it is convenient to use the closely related variables of which s, t, u of (1.2.10) are examples. We define

$$s_{ijk...} = (p_{ijk...})^2, \qquad (4.3.16a)$$

where $$p_{ijk...} = \pm p_i \pm p_j \pm p_k \pm \dots \quad (i \neq j \neq k, \dots). \qquad (4.3.16b)$$

Here the symbols p again refer to the momenta of both the initial and final states, each being preceded by a $(+)$ sign if it refers to an initial-state momentum and by a $(-)$ sign for a final-state momentum.

For a given physical process we classify the variables $(4.3.16a)$ into four categories:

(1) Total energy, referring to that variable which is the square of the total energy in the over-all centre-of-mass system.

(2) Subenergy, referring to those variables which are the square of the energy of a subset of particles in either initial or final state (in their centre-of-mass system).

(3) Momentum transfer, composed of one initial-state and one final-state momentum.

(4) Cross-energy, composed of a mixture of initial-state and final-state momenta, but not (3).

A variable in one category that, because of energy-momentum

conservation, is directly equal to one in another category is usually referred to by the classification that comes higher in the list.

As an example, for the amplitude

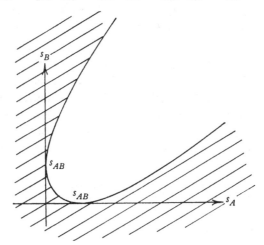

we have one total energy $s_{123} = s_{456}$, six subenergies s_{12}, s_{23}, s_{13}, s_{45}, s_{46}, s_{56}, nine momentum transfers s_{14}, s_{15}, s_{16}, s_{24}, s_{25}, s_{26}, s_{34}, s_{35}, s_{36} and nine cross-energies $s_{145} = s_{236}$, $s_{146} = s_{235}$, $s_{156} = s_{234}$, $s_{245} = s_{136}$, $s_{246} = s_{135}$, $s_{256} = s_{134}$, $s_{345} = s_{126}$, $s_{346} = s_{125}$, $s_{356} = s_{124}$.

Fig. 4.3.1. The shaded region represents the physical region for the two variables s_A and s_B, as given by equation (4.3.20).

Cayley determinants

The Gram-determinant restrictions (4.3.13) and (4.3.15) involve directly the inner-product variables, and it is convenient to change them to forms that directly involve the variables s of (4.3.16a). Following Regge & Barucchi (1964), we do this by introducing Cayley determinants, which are related to the Gram determinants (4.3.10) by the identity

$$G\begin{pmatrix} p_1 \cdots p_N \\ p_1 \cdots p_N \end{pmatrix} = \frac{(-1)^{N+1}}{2^N} \begin{vmatrix} 0 & 1 & 1 & 1 & \cdots & 1 \\ 1 & d_{11} & d_{12} & d_{13} & \cdots & d_{1,N+1} \\ 1 & d_{21} & d_{22} & \cdots & & d_{2,N+1} \\ \cdots & \cdots & \cdots & \cdots & \cdots & \cdots \\ \cdots & \cdots & \cdots & \cdots & \cdots & \cdots \\ 1 & d_{N+1,1} & \cdots & \cdots & d_{N+1,N} & 0 \end{vmatrix}$$

(4.3.17)

where, for $i, j = 1, 2 \ldots N$, $d_{ij} = (p_i - p_j)^2$ and $d_{i,N+1} = d_{N+1,i} = p_i^2$.

The Cayley determinant on the right-hand side of (4.3.17) is not directly useful because the signs in front of p_i, p_j in the definition of d_{ij} are not necessarily in agreement with those in the definition (4.3.16) of s_{ij}. However, it follows from elementary properties of determinants that

$$G\begin{pmatrix} p_1 \cdots p_N \\ p_1 \cdots p_N \end{pmatrix} = G\begin{pmatrix} p_1 p_{12} \cdots p_{12\dots N} \\ p_1 p_{12} \cdots p_{12\dots N} \end{pmatrix}. \qquad (4.3.18)$$

Since, from the definition (4.3.16b),

$$p_{i_1 i_2 \dots i_n} - p_{i_1 i_2 \dots i_r} = p_{i_{r+1} i_{r+2} \dots i_n}, \qquad (4.3.19)$$

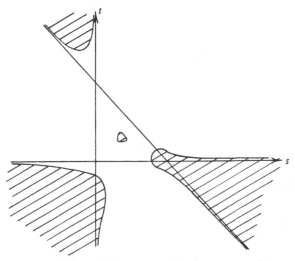

Fig. 4.3.2. The shaded regions are those given by equation (4.3.22). The lower right-hand region is the physical region for the process $1+2 \to 3+4$. The central region corresponds to the decay of one of the particles into the other three, when the masses are suitable. As we see in §4·8, the remaining regions are physical regions for crossed processes.

we see that application of the identity (4.3.17) to the right-hand side of (4.3.18) yields a Cayley determinant whose elements are the variables s of (4.3.16).

We give an illustration of the physical-region condition (4.3.13). Write, for brevity, the set of suffices $i_1 \dots i_r$ as A, $i_{r+1} \dots i_k$ as B and $i_{k+1} \dots i_n$ as C, so that AB denotes $i_1 \dots i_r i_{r+1} \dots i_k$ and so on. Then the condition $G(p_A, p_B) < 0$ becomes, according to (4.3.17) and (4.3.18),

$$\begin{vmatrix} 0 & 1 & 1 & 1 \\ 1 & 0 & s_B & s_A \\ 1 & s_B & 0 & s_{AB} \\ 1 & s_A & s_{AB} & 0 \end{vmatrix} > 0. \qquad (4.3.20a)$$

Multiplied out, this is

$$s_A^2 + s_B^2 + s_{AB}^2 - 2s_A s_B - 2s_A s_{AB} - 2s_B s_{AB} > 0, \qquad (4.3.20b)$$

which for s_{AB} fixed gives the shaded region outside the parabola in Fig. 4.3.1.

Similarly, the condition $G(p_A, p_B, p_C) > 0$ becomes

$$\begin{vmatrix} 0 & 1 & 1 & 1 & 1 \\ 1 & 0 & s_B & s_{BC} & s_A \\ 1 & s_B & 0 & s_C & s_{AB} \\ 1 & s_{BC} & s_C & 0 & s_{ABC} \\ 1 & s_A & s_{AB} & s_{ABC} & 0 \end{vmatrix} > 0. \qquad (4.3.21)$$

Applying this to the amplitude for the process $1 + 2 \to 3 + 4$, letting $A = 1$, $B = 2$, $C = 3$ and putting $s = s_{12}$, $t = s_{23}$, $m_4^2 = s_{123}$, (4.3.21) becomes

$$\begin{vmatrix} 0 & 1 & 1 & 1 & 1 \\ 1 & 0 & m_2^2 & t & m_1^2 \\ 1 & m_2^2 & 0 & m_3^2 & s \\ 1 & t & m_3^2 & 0 & m_4^2 \\ 1 & m_1^2 & s & m_4^2 & 0 \end{vmatrix} > 0. \qquad (4.3.22)$$

This describes the shaded regions in Fig. 4.3.2, first discussed by Kibble (1960).

4.4 Analyticity

The properties that we have introduced so far, unitarity, connectedness, energy-momentum conservation, Lorentz invariance and the connection between spin and statistics, were all discussed by Heisenberg (1943) in his original paper on the S-matrix. But a further crucial property is needed to give a complete theory, that of analyticity. The importance of this was first noted by Kramers [unpublished, see also Heisenberg (1944) and Møller (1946)] but even now, as we noted in §1.1, its precise physical basis is not yet understood.

For this reason we must here present analyticity as a postulate. The crude form of the postulate is that the physical A-matrix elements, or more properly their separate scalar and pseudo-scalar components A_s, A_p as defined in (4.3.9), are boundary values of analytic functions of all the variables s of (4.3.16), these variables now being promoted into complex variables. (The masses of the particles remain fixed at their physical values, and are not regarded as variables.) The statement of the postulate must, however, be refined somewhat to be complete, and that is the main task of this section.

Kinematic singularities

We have seen in § 4.3 that the A-matrix elements may be regarded either as functions of four-momenta or of the scalar variables s of (4.3.16). Equivalent formulations of the analyticity postulate may be given in terms of either set of variables, though we prefer the latter.

In fact, as has been emphasized by Stapp (1963), the equivalence of the two formulations is not transparent owing to technical difficulties concerning points where lower order Gram determinants vanish. This seems related to the possible presence of kinematical singularities in the function A_p of (4.3.9). Which four-momenta should appear in the determinant ϵ in (4.3.9) is a matter of choice, so it would appear that the complete pseudoscalar part ϵA_p of A should generally not vanish for those values of the variables that make ϵ vanish for a particular choice. If this is the case, A_p must have an infinity that cancels the zero of its accompanying ϵ, the position of the infinity depending on the particular choice of momenta included in ϵ.

We must therefore modify the analyticity postulate to allow the possibility of such kinematical singularities. When particles with spin are included in the theory their occurrence may be more widespread (Hearn, 1961); for example the amplitude B in (1.2.21) may be expected to have an infinity that cancels the zero of the spin-factors multiplying it.

Physical region singularities

The A-matrix elements are originally defined in the physical region and can only be continued away from the physical region in an unambiguous way if they are analytic there. We now see that singularities may well occur in the physical region, so that our analyticity assumption must be augmented.

In § 1.3 we saw that 'normal threshold' singularities are likely to arise at points where the physical unitarity equations acquire new terms. This can happen because of new intermediate states in the subenergies and cross-energies as well as the total energy. Examples of terms in equations (4.2.25) which give rise to normal thresholds are

In perturbation theory more complicated physical-region singularities also arise (see § 2.5) and arguments to be given in § 4.10 will suggest that unitarity implies a very similar structure. To illustrate this we outline how this might come about for the triangle singularity (deferring the rigorous discussion until §4.11).

The presence of the last term

$$\text{(diagram)}, \tag{4.4.1}$$

in the unitarity equation in (4.2.25 b) for the amplitude ⊐○⊏ indicates that this amplitude possesses a singularity in the physical region corresponding to the single particle that is common to the two bubbles. We analyse this single-particle singularity in some detail in the next section, where we conclude that, as in perturbation theory, it is actually a pole. Another term in the same unitarity equation takes the form

$$\text{(diagram)}, \tag{4.4.2}$$

and if we insert the single-particle pole in the left-hand bubble we obtain a structure

$$\text{(diagram)} = \text{(diagram)} . \tag{4.4.3a}$$

Here —○— represents the pole factor; according to the rules in (4.2.30) the other internal lines represent δ-functions. Other terms in the unitarity equation are similarly analysed:

$$\text{(diagram)} \longrightarrow \text{(diagram)} = \text{(diagram)} , \tag{4.4.3b}$$

$$\text{(diagram)} \longrightarrow \text{(diagram)} \approx \text{(diagram)} . \tag{4.4.3c}$$

The diagrammatic notation we have adopted suggests that each of the terms in (4.4.3) contains a singularity that coincides in position with that of the triangle Feynman graph in perturbation theory. That our notation does have this interpretation we show generally in § 4.10. The Riemann-sheet properties of the separate terms do not, however, generally agree because Feynman graphs contain poles for all their internal lines.

The presence of the terms (4.4.3) in the unitarity equation for $\equiv\!\!\bigcirc\!\!\equiv$ indicates that the amplitude itself has some sort of triangle singularity in the physical region; in § 4.11 we show that its structure is exactly that of the Feynman graph.

The iε-prescription

The physical-region singularities will divide the physical region into pieces, such that in each piece the amplitudes are analytic. Part of our analyticity postulate will be that, by going round the singularities on paths that go, at least infinitesimally, into the complex s-space, we may analytically continue from any one piece to a neighbouring piece. Equivalently, the physical values of a given amplitude in all the pieces are real-boundary values of the same analytic function.

This immediately raises the question which way round the singularities we must take the paths that link the pieces, or equivalently from which direction in the complex space the real boundary is to be approached. In perturbation theory this question is answered by means of Feynman's rule of attaching a small negative imaginary part $(-i\epsilon)$ to each internal mass, so that the physical-region singularities are displaced out of the real physical region and the pieces are no longer divided from one another.

A normal threshold at the point $(n\mu)^2$ in one of the variables s (with n an integer) is depressed by Feynman's rule into the lower-half plane of that variable. Hence pieces of the physical region to either side of the branch-point are linked by a path passing above the branch-point; the path must be slightly distorted into the upper-half plane to avoid the singularity if now $\epsilon \to 0$. This leads to the rule that we illustrated in Fig. 1.3.4, that normal-threshold cuts are to be approached from above to give the physical boundary value.

We follow convention and adopt the same prescription, called the $+i\epsilon$-prescription, for the normal thresholds outside the context of perturbation theory. The alternatives to so doing would be either to adopt the opposite prescription throughout, or a mixed prescription, one for some of the normal thresholds and the other for the rest. The first of these alternatives would lead to a theory similar to the conventional one, except that conventional amplitudes would be replaced by their complex conjugates. As we indicate in § 4.5, the only physical effect of this would be to introduce a minus sign in the definition of time. The mixed prescription is believed to lead to internal inconsistencies in the theory, as we see below.

We have sketched above, and describe more fully in § 4.10, how the more complicated physical-region singularities are generated from unitarity. The unitarity equations explicitly display the normal thresholds, so that in a sense the more complicated singularities are dependent on the normal thresholds. In particular, our choice of $i\epsilon$-prescription for the normal thresholds fixes that to be adopted for at least some of the more complicated physical-region singularities. This we now show (Olive, 1964; Landshoff & Olive, 1966).

Let such a singularity be given locally by the real equation

$$L(s_i) = 0; \quad [L(s_i^*)^* = L(s_i)]. \tag{4.4.4}$$

Suppose that we wish to decide to which side of some point P on $L = 0$ our path linking pieces of the physical region should pass. It is convenient to confine the part of the path in the neighbourhood of P to the hyperplane (in the multidimensional complex space of the variables s_i) that is normal to $L = 0$ at P. That is, the displacement from P of points on the path takes the form

$$ds_i = \frac{\partial L}{\partial s_i} d\eta, \quad \text{each } i, \tag{4.4.5}$$

the derivatives being evaluated at P. The singularity $L = 0$ occurs at the origin in the complex plane of the variable η so defined—the origin corresponds to the point P. Our problem is to decide whether our path linking the positive and negative real axes in the η-plane should pass above or below the origin.

Once a decision is made one way or the other it is evident that for consistency the same choice must be retained if now the point P is allowed to move around on $L = 0$. In particular, if $L = 0$ touches some other singularity $L' = 0$ we must retain the same prescription at the contact. If we define a variable η' for L', just as η is defined for L in (4.4.5), so that the definitions of η, η' coincide at the contact, the prescription for avoiding L' in the η'-plane is determined at the contact and hence everywhere on L'.

Hence the prescription for any singularity that touches a normal threshold is fixed by our choice of prescription for the normal thresholds, and this in turn fixes the prescription for any other singularity that touches the first one, and so on. While it is not clear that such arguments, and developments of them, cover all physical-region singularities, this may well turn out to be the case. Notice that very often we may find a sequence of singularities, touching one another and including two different normal thresholds. In this case the prescription for one of the normal thresholds fixes that for the other. This is the basis of our earlier statement that a mixed prescription for the normal thresholds, in disagreement with that of perturbation theory, is unlikely to be self-consistent. It may be shown that the perturbation-theory prescription is self-consistent and so, provided S-matrix theory has a physical-region singularity structure similar to that of perturbation theory, no such difficulties arise.†

So far we have only introduced the prescriptions for the physical amplitudes $A^{(+)}$. However, if the amplitude $A_{ab}^{(+)}$ for a transition from state **b** to state **a** is a boundary value of an analytic function, so also presumably the amplitude $A_{ba}^{(+)}$ that describes the reverse transition is a similar boundary value, though not necessarily of the same analytic function. Then $A_{ab}^{(-)} = [A_{ba}^{(+)}]^*$ is the opposite boundary value of an analytic function, because of the elementary theorem that if $f(z)$ is analytic so also is $g(z) \equiv [f(z^*)]^*$ (with $g(z)$ in general different from $f(z)$).

Hence $A_{ab}^{(+)}$, $A_{ab}^{(-)}$ are opposite boundary values of two apparently distinct analytic functions. In § 4.6 we show, however, that these two analytic functions are actually the same. This is the property known as 'hermitian analyticity'.

Normal-threshold structure

As we have suggested in our brief review of the generation of the triangle singularity, the more complicated physical-region singularities may be built up by an iteration procedure. Although this iteration procedure can, in principle, be carried out at this stage of the argument, our description of the subsequent development will be considerably simplified if we ignore all the singularities but the normal thresholds.

† Similar relations between $i\epsilon$-prescriptions follow from any intersections of singularity surfaces with linearly dependent normals, but no such intersections other than the above are known. Our boundary-value postulate also forbids the existence in the physical region of points lying on curves whose local equations do not obey the reality property $L(s_i^*)^* = L(s_i)$, e.g. Landau curves for diagrams involving unstable particles.

At present this procedure does, indeed, seem to be in part necessary, for we shall in subsequent sections have to make arguments concerning certain unphysical regions lying at real points outside the physical region. Amongst the singularities expected to lie in such regions are those corresponding to diagrams with three-line vertices. Such a vertex is 'unphysical' when it involves only stable particles and so cannot possibly be analysed at the present stage of the argument. An example of a singularity in which such vertices occur is the triangle anomalous threshold in two-particle → two-particle scattering (see Fig. 2.4.3a). So the procedure that apparently must be adopted is to start with the normal thresholds, prove the fundamental theorems of subsequent sections, and then use the results of these to generate the 'unphysical' singularities by iteration (Polkinghorne, 1962a,b; Olive, 1964). As each new such singularity is discovered in the iteration procedure, it should be checked that its presence does not upset any of the fundamental theorems.

The structure of the normal thresholds itself is quite complicated, in that for a multiparticle amplitude physical-region normal thresholds $s = (n\mu)^2$ may be associated with many variables s. In fact, with the terminology of the classification in § 4.3, they will occur in all the variables s except the momentum transfers, that is in the total energy, the subenergies and the cross-energies.

Now we saw in § 4.3 that various constraints relate the variables s. If we were to apply these constraints and so work with an independent subset of the variables, the normal thresholds corresponding to the redundant variables would, in a complicated way and with complicated $i\epsilon$-prescriptions, appear in the space of the independent variables. We therefore prefer to retain all the variables and rather regard the constraints as auxiliary conditions that restrict the variables to lie on certain surfaces in the larger space. By so doing we retain an important piece of information, whose truth would not otherwise be so evident. This is that normal thresholds corresponding to different variables s are independent. By this we mean that if we go from a point P round a normal threshold corresponding to a variable s_1, then round one corresponding to s_2, and back to P, we get the same result for the final value of the function as if we encircled the two thresholds in the reverse order (though in the same sense as before). If we allow the presence of other singularities this statement must, of course, be interpreted with some care. Its truth may be established by showing that the two paths may be continuously distorted into each other; the easiest way to convince oneself of this is to realise that their projections on to the complex s_1 and s_2 planes are identical for the two cases. In this argument it is crucial that the two singularities in question do not touch each other.

4.5 Physical-region poles

We here analyse the simplest singularities, those associated with single stable particles. In perturbation theory these singularities are simple poles, as represented for example by the Feynman graph of Fig. 1.5.3. We now show that this is the case also outside the framework of perturbation theory (Olive, 1964).

The argument is simplest for the poles that occur in the physical region of an amplitude, for there we may make use directly of the physical unitarity equations. The simplest case arises from the three-particle unitarity equation of (4.2.25 b):

$$(4.5.1)$$

According to the rules (4.2.30) for interpreting diagrammatic equations, each of the terms in the last sum in (4.5.1) contains a δ-function; for example

$= A_{22}^{(+)}[-2\pi i\, \delta(q^2-\mu^2)]\, A_{22}^{(-)}.$ $(4.5.2)$

Here q is the four-momentum carried by the internal line,

$$q = (p_1 + p_2 - p_4),$$

so that by the classification of §4.3 q^2 is a cross-energy variable. $A_{22}^{(+)}$ and $A_{22}^{(-)}$ are the amplitudes for the two bubbles contained in (4.5.2).

Evidently (4.5.2) is infinite at $q^2 = \mu^2$, and so some other term or terms in (4.5.1) must also be infinite to balance this. As will become clearer from the analysis of §4.10, none of the *integrations* in the other terms of (4.5.1) can produce an infinity at a value of q^2 independent of the other variables, and it must be the amplitude ⊐○⊏ that contains the infinity. This is evident if it is accepted that all singularities are associated with diagrams and have properties at least reminiscent of those in the corresponding Feynman graphs.

An infinity of an amplitude in the physical region must surely be capable of physical interpretation, and here such an interpretation is not difficult to find. Our theory is formulated in terms of momentum eigenstates, which contain no information as to the positions of the particles. Thus the lines of flight of the particles are overwhelmingly likely to be such that that particles 'miss' one another, or at most interact in groups. This idea led to the connectedness structure of § 4.2. The next most likely possibility is that first one set of particles interacts, and that one of the particles emerging from this interaction at a later time interacts with another set of particles. This is exactly what happens in an experimental situation; particles are produced in an accelerator and then at a later time and in a distant place (distant compared with the range of the fundamental forces) are made to suffer the interaction that is the object of the experiment. The existence of the infinity in the amplitude that describes the combined events implies that when one sets up the apparatus to try and achieve the two successive events, the likelihood of their happening is infinite compared with that of all the particles involved interacting simultaneously.

These considerations lead us to suppose that the amplitude contains a part

$$\text{} = A_{22}^{(+)} D^{(+)}(q^2) A_{22}^{(+)}, \qquad (4.5.3a)$$

so that contains the hermitian-conjugate part

$$\text{} = A_{22}^{(-)} D^{(-)}(q^2) A_{22}^{(-)}. \qquad (4.5.3b)$$

Here
$$\text{} = D^{(+)}(q^2), \qquad (4.5.4a)$$

and
$$\text{} = D^{(-)}(q^2) = [D^{(+)}(q^2)]^*. \qquad (4.5.4b)$$

The function $D^{(+)}(q^2)$ represents the propagation of the particle in the free flight between successive scatterings and is infinite when that particle is on the mass shell. We aim to show that, as in perturbation theory, the function $D^{(+)}(q^2)$ is a simple pole.

We insert the behaviour (4.5.3) into the unitarity equation (4.5.1) and pick out those terms that are infinite at $q^2 = \mu^2$. This gives

$$(4.5.5)$$

To arrive at this equation we have again used the ideas described more fully in §4.10 to deduce that the first two terms on the right-hand side of (4.5.1) involve integrations in such a way that these terms do not contain a singularity at a fixed value of q^2. The equation (4.5.5) may be simplified by use of the two-particle unitarity equation (4.2.23). Thus

$$(4.5.6)$$

with similar manipulations for the second term on the right-hand side of (4.5.5). In this way we obtain

$$(4.5.7a)$$

or $\quad A_{22}^{(+)} D^{(+)} A_{22}^{(-)} - A_{22}^{(+)} D^{(-)} A_{22}^{(-)} = A_{22}^{(+)} [-2\pi i \delta(q^2 - \mu^2)] A_{22}^{(-)}.$

$$(4.5.7b)$$

A trivial cancellation then gives

$$D^{(+)} - D^{(-)} = -2\pi i\, \delta(q^2 - \mu^2). \tag{4.5.8}$$

The general solution of (4.5.8) and (4.5.4b) is

$$D^{(+)}(q^2) = \frac{\lambda}{q^2 - \mu^2 + i\epsilon} - \frac{1-\lambda}{q^2 - \mu^2 - i\epsilon}, \tag{4.5.9}$$

where λ is any real number. (We could add in to this general solution a function regular at $q^2 = \mu^2$, but this would merely have the effect of changing the definition of the background term remaining when (4.5.3a) is subtracted from $\equiv\!\!\bigcirc\!\!\!+\!\!\!\equiv$).

Choice of λ

It remains to determine the value of λ in (4.5.9), and there is more than one way in which we may do this. First we notice that any value of λ other than 0 or 1 would make $D(q^2)$ non-analytic, and so can be excluded by the simple assumption that $\equiv\!\!\bigcirc\!\!\equiv$ is to be analytic. This assumption was part of our analyticity postulate. We then fix the conventional value $\lambda = 1$ by saying that the single-particle poles are to be included in the $+i\epsilon$-prescription introduced in §4.4.

This approach makes no direct mention why *physics* should require $\lambda = 1$. Presumably it is physics that requires $\equiv\!\!\bigcirc\!\!\equiv$ to be analytic, but we have said that, as yet, we have no satisfactory way of demonstrating this. We should therefore prefer to understand directly why $\lambda = 1$.

Our physical interpretation of the single-particle singularity is, as we have said, that of two *successive* scatterings. To show that this requires $\lambda = 1$ we must somehow introduce into the theory the notion of the direction of time. The crude way to do this (Branson, 1964) is to define, in terms of an A-matrix element, a function

$$\tilde{A}(\tau) = \frac{1}{(2\pi)^{\frac{1}{2}}} \int_{-\infty}^{\infty} dE\, e^{-iE\tau} A(E). \qquad (4.5.10)$$

Here E is the total energy and other variables are suppressed. We want to interpret $|\tilde{A}(\tau)|^2 d\tau$ as a relative probability that the time duration of the interaction described by A lies between τ and $\tau + d\tau$.

Two things make this interpretation plausible. One is that time and energy are conjugate variables in the sense of Fourier transformation, in ordinary quantum mechanics. The other is the convolution property of Fourier integrals, which says that the Fourier transform of (4.5.3a) is

$$\tilde{A}(\tau) = \int_{-\infty}^{\infty} d\tau_1 d\tau_2 d\tau_3\, \delta(\tau - \tau_1 - \tau_2 - \tau_3)\, \tilde{A}_{22}(\tau_1)\, \tilde{D}(\tau_2)\, \tilde{A}_{22}(\tau_3), \quad (4.5.11)$$

where $\tilde{D}$ has a definition similar to (4.5.10). Because of the δ-function we interpret this as saying that the probability amplitude for the overall time of interaction being τ is the integral over the amplitudes for the individual scatterings, together with the intermediate free flight, taking a total time τ.

If we accept this interpretation, the requirement that the free-flight time be positive is $\tilde{D}(\tau) = 0$ for $\tau < 0$.   (4.5.12)

Because of the inversion property of Fourier integrals

$$D(E) = \frac{1}{(2\pi)^{\frac{1}{2}}} \int_{-\infty}^{\infty} e^{iE\tau} \tilde{D}(\tau) \, d\tau,   (4.5.13)$$

this means that $D(E)$ is analytic in the upper half E-plane. (If the integral (4.5.13) converges for real E, the exponential factor will enhance the convergence for $\text{Im } E > 0$ if the integration only involves $\tau > 0$.) By taking a Lorentz frame in which E is simply related to q^2 we see that D must also be analytic in the upper half q^2-plane, so that $\lambda = 1$.

Normal thresholds may be treated similarly (Branson, 1964), thus giving a simple physical interpretation of the $i\epsilon$-prescription for them. Notice that we might well have chosen the opposite sign in the exponential in the definition (4.5.10), that is we might have defined time to occur with the opposite sign. The result then would be $\lambda = 0$ and the opposite $i\epsilon$-prescription for the normal thresholds.

The above discussion has many unsatisfactory features. For example, the integral in (4.5.10) extends to negative energies, and it surely cannot be that the definition of physical time involves unphysical energies. This means that a precise localisation in time is not really possible and, contrary to the implication of the requirement (4.5.12), time is at best a macroscopic quantity. The microscopic condition (4.5.12) must essentially be replaced by the requirement that the amplitude for the time of flight $\to 0$ as $\tau \to -\infty$. For a discussion of how this still results in the value $\lambda = 1$, the reader is referred to papers by Wanders (1965), Stapp (1964b), Iagolnitzer (1965), Peres (1965). The physics behind the idea of macroscopic time is discussed by Eden & Landshoff (1965).

There is another way of incorporating the requirement of positive flight times (Landshoff & Olive, 1966). We saw in § 4.4 that when the single-particle singularities are inserted in unitarity they generate more complicated physical-region singularities. We show in § 4.11 that, if we suppose that the single-particle singularity does indeed have $\lambda = 1$, the triangle singularity has exactly the same physical-region structure as in perturbation theory; this is probably also the case for the other singularities. But in § 2.5 we saw that the physical-region singularities of perturbation theory can be given the physical interpretation of representing a succession of interactions having physical momenta and physical (that is positive) intermediate flight times. Any other value of λ would certainly result in some of the more complicated singularities occurring in parts of the physical region where they cannot be given this interpretation; one would have to say that some of the internal particles were moving backwards in time.

The single-particle discontinuity

We gave a physical argument for the factorisation (4.5.3) of the single-particle singularity into a product of scattering amplitudes and a 'propagator' D for the intermediate particle. We then showed that this is consistent with unitarity. We now show that it is actually uniquely demanded by unitarity. This may be done by an extension of our previous methods, but we present instead an argument that involves calculating the 'discontinuity' across the single-particle singularity.

This new argument eliminates one inelegance of the previous one. This is that the amplitudes A_{22} appearing in (4.5.3) are only defined at $q^2 = \mu^2$. Away from this value of q^2 the theory provides no unique definition of A_{22}, and we may use any one that is analytic. Different definitions merely lead to different definitions of the background term remaining when (4.5.3) is subtracted from the amplitude $\equiv\!\!\bigcirc\!\!\equiv$.

In the unitarity equation (4.5.1) the term (4.5.2) contributes only at $q^2 = \mu^2$. Hence the equation changes non-analytically at this value of q^2. If we take the form of the equation that is valid for $q^2 < \mu^2$ and continue it analytically to $q^2 = \mu^2$, the extra term (4.5.2) does not appear. Hence we have two equations valid at $q^2 = \mu^2$, one containing (4.5.2) and one not.

Suppose that the continuation we choose to make is that which takes q^2 just below μ^2, $q^2 \to \mu^2 - i\epsilon$. Because of our $i\epsilon$-prescription of § 4.4, this results in $\equiv\!\!(-)\!\!\equiv \to \equiv\!\!(-)\!\!\equiv$, but for $\equiv\!\!(+)\!\!\equiv$ we have gone to the wrong side of the singularity, so we write $\equiv\!\!(+)\!\!\equiv \to \equiv\!\!(i)\!\!\equiv$. As for the other terms in (4.5.1) that contain the singularity

$$\overrightarrow{}\!\!\!(+)\!\!-\!\!(-)\!\!\! \longrightarrow \overrightarrow{}\!\!\!(+)\!\!-\!\!(-)\!\!\! \quad \text{and} \quad \equiv\!\!(+)\!\!=\!\!\ominus\!\!\! \longrightarrow \equiv\!\!(i)\!\!=\!\!\ominus\!\!\!, \qquad (4.5.14)$$

because in the latter term q^2 is not being integrated over. So as a whole the continuation of (4.5.1) reads

$$\equiv\!\!(i)\!\!\equiv - \equiv\!\!(-)\!\!\equiv = \overrightarrow{}\!\!\!(+)\!\!-\!\!(-)\!\!\! + \equiv\!\!(i)\!\!=\!\!\ominus\!\!\! + R, \qquad (4.5.15)$$

where R does not contain the one-particle singularity. Subtracting this from (4.5.1), that is from

$$+ \; R, \tag{4.5.16}$$

we obtain

or, rearranging,

$$\tag{4.5.17}$$

If we 'postmultiply' by

$$\tag{4.5.18}$$

and use the unitarity equation

$$\tag{4.5.19a}$$

so that

$$\tag{4.5.19b}$$

we obtain

$$\tag{4.5.20}$$

and this is the discontinuity we are seeking. To show that this again implies the existence of a pole in ⊒○⊑ we must make arguments similar to those following (4.5.8).

Notice that this analysis shows that the pole must be simple;[†] a pole of order n would require a discontinuity proportional to

$$\left(\frac{d}{dq^2}\right)^{n-1} \delta(q^2 - \mu^2).$$

[†] This argument only applies to real physical-region poles. A mechanism for the possible existence of complex higher-order poles has been described by Eden & Landshoff (1964). See also Goldberger & Watson (1964b).

Further, its existence depends on the existence of a corresponding physical stable particle. If there were no such particle the term (4.5.2) would be totally absent and the calculation of the discontinuity would just yield the result zero. So there is a one-one correspondence between physical-region poles and stable particles.

Generalisations

So far we have only considered the poles of the amplitude in that part of the physical region where the unitarity equation (4.5.1) applies. At higher energies more terms enter the unitarity equation, but the same result (4.5.3) must apply, as may be seen by writing it in the form

$$[(q^2 - \mu^2) A_{33}^{(+)}]_{q^2 = \mu^2} = A_{22}^{(+)} A_{22}^{(+)},$$

and continuing analytically along $q^2 = \mu^2$ up to higher energies. We also expect similar properties for the higher amplitudes, that is those involving more particles. For example

$$(4.5.21)$$

In fact the extension of the validity of (4.5.3) to higher energies and the generalisations to higher amplitudes are inter-related, because of the way the amplitudes are coupled by unitarity. This we now demonstrate, by considering the energies at which the unitarity equations (4.2.25c) apply, that is $(4\mu)^2 \leqslant (\Sigma p)^2 < (5\mu)^2$. In this range of energy the unitarity equation for reads

$$(4.5.22)$$

Noting that the term (4.5.2) in this equation only contributes at $q^2 = \mu^2$, and proceeding exactly as before we obtain

$$(4.5.23)$$

Subtraction of this from (4.5.22) and rearrangement yields

$$(4.5.24)$$

The same treatment for the corresponding unitarity equation for ⊐◯⊏ gives

$$(4.5.25)$$

The equations (4.5.24) and (4.5.25) are two simultaneous equations for the discontinuities of ⊐◯⊏ and ⊐◯⊏. To solve for the former, postmultiply (4.5.24) by (4.5.18) and (4.5.25) by

The unitarity equations

$$(4.5.26a)$$

$$(4.5.26b)$$

then lead to the same result as in (4.5.20). Similarly, post-multiplication of (4.5.24) by

and of (4.5.25) by

leads to the result

$$(4.5.27)$$

which is in agreement with the structure in (4.5.21).

This example leads us to expect that the physical-region pole factorisation can, in principle, be proved for all multi-particle amplitudes. But it is evident that the argument becomes enormously complicated at higher energies, as more and more amplitudes become coupled.

With the higher amplitudes a new feature appears, that of twofold poles (or more). For example, inserting (4.5.3) into the structure

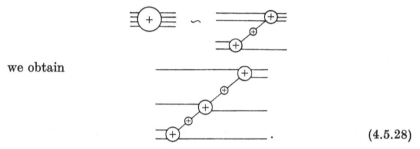

we obtain

$$\tag{4.5.28}$$

4.6   Hermitian analyticity and extended unitarity

Having analysed the single-particle discontinuity, we now study the discontinuities associated with the normal thresholds in the physical region (Olive, 1963a, b; Gunson, 1965).

According to the unitarity equations (4.2.25) the amplitude ⊃◯⊂ satisfies, in the energy range $(2\mu)^2 \leqslant (\Sigma p)^2 < (3\mu)^2$,

$$⊃⊕⊂ - ⊃⊖⊂ = ⊃⊕⊂⊃⊖⊂ = ⊃⊖⊂⊃⊕⊂ \tag{4.6.1}$$

and in $(3\mu)^2 \leqslant (\Sigma p)^2 < (4\mu)^2$

$$⊃⊕⊂ - ⊃⊖⊂ = ⊃⊕⊂⊃⊖⊂ + ⊃⊕⊂⊃⊖⊂$$

$$= ⊃⊖⊂⊃⊕⊂ + ⊃⊖⊂⊃⊕⊂. \tag{4.6.2}$$

Later in this section we prove a property known as 'hermitian analyticity', a consequence of which is that (4.6.1) actually represents the discontinuity of ⊃◯⊂ associated with the two-particle normal threshold, while (4.6.2) represents that associated with the two-particle and three-particle thresholds combined. Even without knowing this, however, we may use (4.6.1) and (4.6.2) to determine the discontinuity associated with the three-particle threshold alone.

We continue the equation (4.6.1) analytically in the variable $s = (\Sigma p)^2$, into the region where (4.6.2) operates, and then compare the two equations. In the continuation we must take a detour round the branch-point at $s = (3\mu)^2$, and it must be the same detour for every term in the equation. We choose a detour that goes below the branch-point. Then according to the $i\epsilon$-prescription of §4.4,

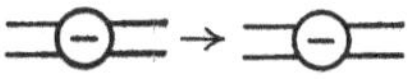

$$\boxed{(-)} \rightarrow \boxed{(-)}$$

(see Fig. 4.6.1 a), but for 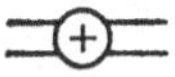 $\boxed{(+)}$ we are going to the 'wrong' side of the $(3\mu)^2$ threshold, and so we write

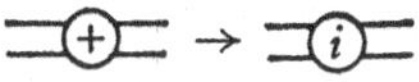

$$\boxed{(+)} \rightarrow \boxed{(i)}$$

(see Fig. 4.6.1 b). So the continuation of (4.6.1) into $(3\mu)^2 \leqslant s < (4\mu)^2$ is

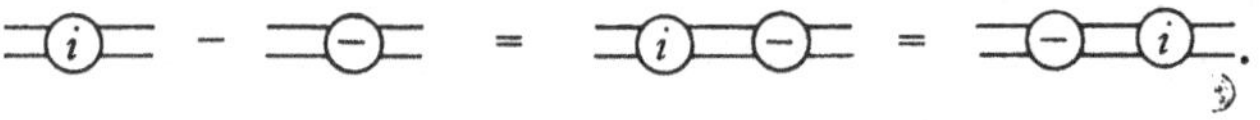

$$\boxed{(i)} - \boxed{(-)} \;=\; \boxed{(i)}\boxed{(-)} \;=\; \boxed{(-)}\boxed{(i)}. \qquad (4.6.3)$$

Fig. 4.6.1. The path of continuation applied to equation (4.6.1.), (a) shows the path for the $(-)$ amplitude, and (b) the corresponding path for the $(+)$ amplitude.

Introduce now a new amplitude $\boxed{(i)}$, defined by the equation

$$\boxed{(i)} - \boxed{(-)} \;=\; \boxed{(i)}\boxed{(-)}. \qquad (4.6.4)$$

If we 'premultiply' (4.6.4) by  $\boxed{(-)}$ and use (4.6.3) we see that $\boxed{(i)}$ also satisfies

$$\boxed{(i)} - \boxed{(-)} \;=\; \boxed{(-)}\boxed{(i)}. \qquad (4.6.5)$$

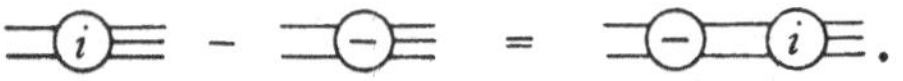

Now we can say that the following expression vanishes, because each bracket vanishes

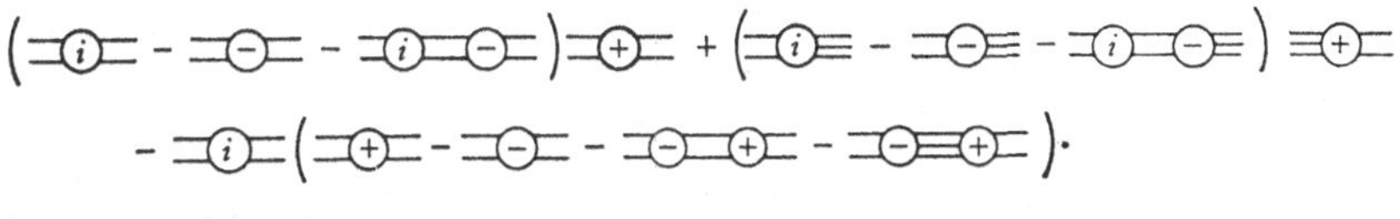

$$\left(\boxed{(i)} - \boxed{(-)} - \boxed{(i)}\boxed{(-)}\right)\boxed{(+)} + \left(\boxed{(i)} - \boxed{(-)} - \boxed{(i)}\boxed{(-)}\right)\boxed{(+)}$$
$$- \boxed{(i)}\left(\boxed{(+)} - \boxed{(-)} - \boxed{(-)}\boxed{(+)} - \boxed{(-)}\boxed{(+)}\right).$$

Several terms here cancel, so that

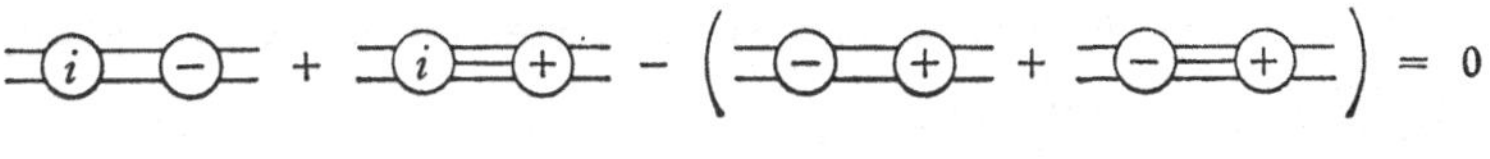

$$\boxed{(i)}\boxed{(-)} + \boxed{(i)}\boxed{(+)} - \left(\boxed{(-)}\boxed{(+)} + \boxed{(-)}\boxed{(+)}\right) = 0$$

Using (4.6.3) and (4.6.2) we finally obtain from this

$$\boxed{\text{—(+)—}} \; - \; \boxed{\text{—(}i\text{)—}} \;\; = \;\; \boxed{\text{—(}i\text{)==(+)—}}, \qquad (4.6.6)$$

which is the desired discontinuity of —(+)— across the three-particle cut expressed as a single integral involving the three-particle intermediate state. Note that it does not take the form of the extra term in (4.6.2) as compared with (4.6.1), because of the label (i) replacing the $(-)$. The important feature of the argument, which can be generalised to subenergy and cross-energy variables, is that the discontinuity around a normal threshold occurring inside the physical region is derived from a comparison of the physical unitarity equation operating on either side of the threshold.

Notice that, although we have defined —(i)= by (4.6.4), the resemblance of (4.6.4) to (4.6.3) suggests that —(i)= may be obtained as a continuation of —(+)=, just as —(i)— is a continuation of —(+)—. If this is the case, the equations (4.6.4) and (4.6.5), valid in $(3\mu)^2 \leqslant s < (4\mu)^2$, will have as continuations into $(2\mu)^2 \leqslant s < (3\mu)^2$

$$\text{—(+)=} \; - \; \text{—(−)=} \;\; = \;\; \text{—(+)—(−)=} \;\; = \;\; \text{—(−)—(+)=}. \qquad (4.6.7)$$

Here the energy is such that the intermediate two-particle state is physical, though for the initial state both the total energy and the three subenergies are below their physical thresholds. Such a relation cannot be a physical unitarity relation since it relates continuations of the amplitude to points outside the physical region. It is known as an 'extended unitarity' relation and its proof is discussed later in this section. Further examples of extended unitarity relations, valid in the same energy range $(2\mu)^2 \leqslant s < (3\mu)^2$, are

$$\text{=(+)—} \; - \; \text{=(−)—} \;\; = \;\; \text{=(+)—(−)—} \;\; = \;\; \text{=(−)—(+)—}, \qquad (4.6.8a)$$

$$\text{=(+)=} \; - \; \text{=(−)=} \;\; = \;\; \text{=(+)—(−)=} \;\; = \;\; \text{=(−)—(+)=}. \qquad (4.6.8b)$$

Just as hermitian analyticity will result in (4.6.1) being the discontinuity of —O— across the two-particle cut in the total-energy variable s, so are (4.6.7) and (4.6.8) the corresponding discontinuities of —O=, =O— and =O=.

Hermitian analyticity

When the energy increases, the physical unitarity equations acquire extra terms. When it decreases they shed terms; this is illustrated by equations (4.6.1) and (4.6.2). Likewise we have said that, when s is taken below $(3\mu)^2$, we expect that the physical unitarity equation

$$\boxed{+} - \boxed{-} = \boxed{+}\,\boxed{-} + \boxed{+}\boxminus\boxminus\boxminus\boxminus + \Sigma\,\boxed{+}\boxminus\boxminus\boxminus \qquad (4.6.9)$$

is replaced by the extended unitarity equation (4.6.7). In the same way we might expect that, for $s < (2\mu)^2$, the physical unitarity equation (4.6.1) is replaced by

$$\boxed{+} - \boxed{-} = 0. \qquad (4.6.10)$$

This relation, if true, states that specific continuations out of the physical region of the amplitudes $\boxed{+}$ and $\boxed{-}$, indicated in Fig. 4.6.2, are equal. So while in §4.4 we said that, in the physical region, $\boxed{+}$ and $\boxed{-}$ are boundary values of analytic functions that are apparently unrelated, we now say that they are actually opposite boundary values of the same analytic function. This property is known as 'hermitian analyticity'.

Fig. 4.6.2. The path of continuation applied to the ($+$) and ($-$) amplitudes respectively to achieve the hermitian analyticity condition (4.6.10).

An immediate and vital consequence of hermitian analyticity is that unitarity equations evaluate discontinuities (Olive, 1962; Stapp, 1962b). For example (4.6.1) gives the discontinuity of $\boxed{\bigcirc}$ across the two-particle cut.

Proofs of the hermitian analyticity property have been given in perturbation theory (Olive, 1962) and in potential theory (J. R. Taylor, 1964). The proof we outline here is based on S-matrix theory alone. It was first given by Olive (1964) and discussed in greater detail by Boyling (1964b). Although we give the proof for the particular case of the equation (4.6.10) we expect that very similar arguments could be applied to give the extended unitarity relations, for example (4.6.7) and (4.6.8). This is because both hermitian analyticity and extended unitarity can be regarded as illustrations of a more general principle, which we might call 'unphysical unitarity', whereby unitarity-like formulae operates outside the physical region with the intermediate states still determined by the energy range.

Four-particle unitarity

The derivation of the hermitian analyticity property (4.6.10) makes use of the single-particle pole structure analysed in §4.5 and of the physical unitarity equations for the amplitude ⟩⟨.

According to (4.5.28) this amplitude contains a twofold pole in the physical region:

$$(4.6.11a)$$

so that, by hermitian conjugation,

$$(4.6.11b)$$

In §4.5 we deduced the structure (4.6.11) by an indirect argument; here it is useful to check it directly by means of the unitarity equation for ⟩⟨ that operates below the five-particle threshold. From (4.2.25c) this reads

$$+ R. \qquad (4.6.12)$$

Here, as a result of the ideas of §4.10 that all singularities are represented by diagrams and have properties something like those of the corresponding Feynman graphs, the function R is free of the twofold singularity under examination. We insert (4.6.11), together with the single-pole structure (4.5.3) for the amplitude ⊐○⊏, into (4.6.12) and pick out the structure of the twofold singularity in each term:

$$(4.6.13)$$

Contained within the first and second terms of this equation is the part ⊐(+)⊐(−)⊏ , which may be reduced by means of the unitarity equation (4.6.1). If we also use the relation (4.5.8),

$$—(+)—\ -\ —(−)—\ =\ ———\ ,\qquad (4.6.14)$$

we find that various cancellations occur, and (4.6.13) reduces to

$$(4.6.15)$$

The outside bubbles and the propagators are common factors of each term in this equation. If we cancel them, we are left with (4.6.1) and so have verified the consistency of (4.6.11).

    The point of the analysis above is that it now suggests how to obtain the hermitian analyticity (4.6.10), which is simply (4.6.1) with zero right-hand side.

Derivation of hermitian analyticity

If we trace back through the foregoing argument, we find that it is the term

$$(4.6.16)$$

in (4.6.12) that results in the right-hand side of (4.6.15). This term does not contribute to the unitarity equation (4.6.12) when

$$(p_1 + p_2 + p_3 - p_5)^2 \equiv s_{1235} < (2\mu)^2. \tag{4.6.17}$$

We shall only be interested in positive values of

$$s_{125} \equiv (p_1 + p_2 - p_5)^2$$

and it then follows from (4.3.8) that

$$(p_1 + p_2 - p_5) \cdot p_3 \geqslant \sqrt{(s_{125})}\,\mu \tag{4.6.18}$$

or rearranging $s_{1235} \geqslant (\sqrt{s_{125}} + \mu)^2.$

Taking the square root $\sqrt{s_{1235}} \geqslant \sqrt{s_{125}} + \mu. \tag{4.6.19a}$

Likewise $\sqrt{s_{1235}} \geqslant \sqrt{s_{478}} + \mu, \tag{4.6.19b}$

where $s_{478} \equiv (p_4 - p_7 - p_8)^2.$

When (4.6.17) holds it follows from (4.6.19) that

$$s_{125} < \mu^2, \quad s_{478} < \mu^2. \tag{4.6.20}$$

The third and fourth terms on the right-hand side of (4.6.12) contribute only when $s_{125} = \mu^2$ and $s_{478} = \mu^2$. Hence when (4.6.17) holds the unitarity relation (4.6.12) is replaced by

$\tag{4.6.21}$

Of the many variables the most important ones for the present argument are s_{1235}, s_{125} and s_{478}. s_{125} and s_{478} play equivalent roles, and we can equate them and picture the situation two-dimensionally by plotting this joint variable against s_{1235} as in Fig. 4.6.3.

According to (4.6.19) the physical region in terms of these variables is given by an arc of a parabola passing through the intersection of $s_{1235} = (2\mu)^2$ and $s_{125} = s_{478} = \mu^2$. (Rationalising (4.6.19a) and putting $A = 125$, $B = 3$ we have in fact (4.3.20), plotted in Fig. 4.3.1.) We shall call the regions in which (4.6.12) and (4.6.21) operate R_1 and R_2 respectively, adding the extra labels $\pm$ according to whether we have in mind the physical regions for the $+$ or $-$ amplitudes.

Equation (4.6.20) means that the pole structures (4.6.11) do not occur in R_2 so that if we are to use (4.6.21) in the same way as (4.6.12) we must continue it out of R_2. According to our analyticity postulate the $(-)$ amplitude can be continued from R_2^- to R_1^- by a path within

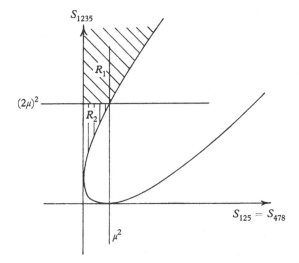

Fig. 4.6.3. The parts of the physical region for the four-particle $\rightarrow$ four-particle amplitude. In R_1 the term (4.6.16) is present in the unitarity equation; in R_2 it is not. The diagram shows the section of the space of the variables on which $s_{125} = s_{478}$. The hatched region is the physical region.

the physical region. If we continue equation (4.6.21) so that the $(-)$-amplitude follows this path the $(+)$-amplitude will follow a related path leading not to R_2^+ but to a new region which we call R_1^i.

We know that the $(-)$-amplitude possesses the physical region pole structure (4.6.11b) in R_1^-, but only know that the $(+)$-amplitude possesses the structure (4.6.11a) in R_1^+ but not necessarily in R_1^i. If we can find a path of continuation relating R_1^+ to R_1^i and lying on the mass shell section $s_{125} = s_{478} = \mu^2$, we can continue the relation (4.6.11a) to R_1^i. Such a path can be found if the path we already have, from R_1^+ to R_2^+ to R_1^i, can be distorted to lie on the mass shell section without cutting any singular surfaces.

This question cannot be answered finally without complete knowledge of the singularity structure but in terms of the 'approximation' explained in §4.4 whereby we retain only normal thresholds the question is easily settled. The situation can be pictured as in Fig. 4.6.4 where we have drawn the known path from R_1^+ to R_2^+ to R_1^i running over the $s_{1235} = (2\mu)^2$ threshold with a $+i\epsilon$-distortion and returning with the opposite distortion. This path can obviously be distorted to lie on the mass shell section $s_{125} = s_{478} = \mu^2$, as indicated by the arrows. (Note that, since it is only a neighbourhood of the point of

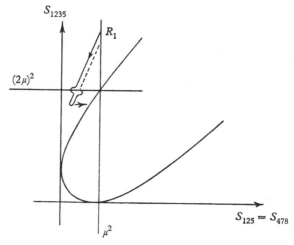

Fig. 4.6.4. The path with arrows runs from R_1^+ to R_2^+ to R_1^i. The large arrows indicate the deformation of this path through the unphysical region bounded by the parabola to the mass-shell section $s_{125} = s_{478} = \mu^2$.

intersection of $s_{1235} = (2\mu)^2$ and $s_{125} = s_{478} = \mu^2$ that matters, the argument would be rigorous if we knew no other singularity curves passed through this point. In fact no such Landau curves are known (Boyling, 1964 *b*).)

Thus continuing (4.6.11 *a*) from R_1^+ to R_1^i by a path looping just the $s_{1235} = (2\mu)^2$ threshold we find

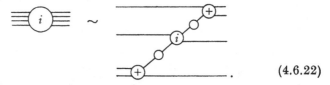

$$(4.6.22)$$

The label (i) on the middle bubble indicates that it is evaluated in the region reached when its energy variable s_{1235} encircles $(2\mu)^2$ in an anticlockwise sense.

Inserting (4.6.22) and (4.6.11b) into the continuation of (4.6.21) and picking out the parts with the twofold pole structure we have

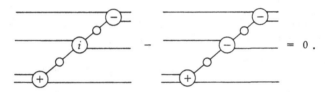

Notice that there is no need to label ——○—— with a (+) or a (−) because the continuation is being made to the poles rather than above or below them. Using (4.6.1) to rewrite the terms on the right-hand side as we did before we find, on cancelling,

Cancelling off the outer bubbles and ——○—— we see that

This is the desired hermitian analyticity relation stating that the (−)-amplitude equals the continuation of the (+)-amplitude around the $(2\mu)^2$ threshold.

The modification of the above argument required when anomalous thresholds are present is discussed in §4.11.

4.7 Normal-threshold discontinuities

We have been considering a singularity scheme in which we discuss only normal thresholds and this must be supplemented by expressions for the discontinuities across the associated cuts. The discontinuities corresponding to intermediate states explicitly appearing in the physical unitarity relation (i.e. those in energy, subenergy and cross-energy variables, but not momentum transfer variables) could, in principle, be deduced by the methods of the previous section, but this is likely to be tedious. (A start has been made by Boyling (1964).) Instead we shall temporarily work backwards by suggesting formulae for such discontinuities and showing that they are consistent with the ideas so far developed (Olive, 1965; see also Stapp, 1965). First, however, we discuss the nature of the two-particle thresholds.

Nature of the normal threshold branch-points

We saw that the discontinuity across the $(2\mu)^2$ cut is given by the two-particle unitarity equation

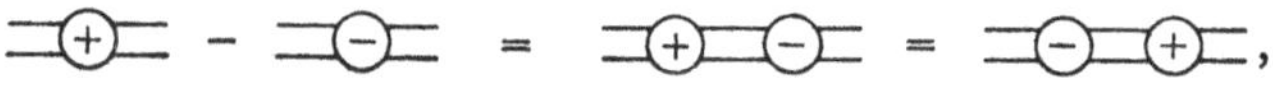

$$ \text{—}(+)\text{—} \;-\; \text{—}(-)\text{—} \;=\; \text{—}(+)\text{—}(-)\text{—} \;=\; \text{—}(-)\text{—}(+)\text{—}, \qquad (4.7.1) $$

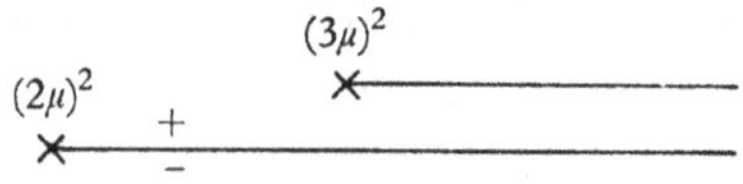

Fig. 4.7.1. $(+)$ and $(-)$ boundary values in the region $(2\mu)^2 < s < (3\mu)^2$.

where $+$ and $-$ now refer to the boundary values in Fig. 4.7.1. According to the rule (4.2.30) this can be written

$$ A^{(+)} - A^{(-)} = \frac{1}{2} \int \frac{i d^4 k}{(2\pi)^4} (-2\pi i)^2 \, \delta^{(+)}(q_1^2 - \mu^2) \, \delta^{(+)}(q_2^2 - \mu^2) \, A^{(+)} A^{(-)} $$

$$ = \frac{-i \sqrt{\{s - (2\mu)^2\}}}{(8\pi)^2 \sqrt{s}} \int d\Gamma \, A^{(+)} A^{(-)}, \qquad (4.7.2) $$

when worked out in the centre-of-mass system, where $\int d\Gamma$ is the angular integration for the centre-of-mass momentum of the particles in the intermediate state.

If we continue (4.7.2) around $s = (2\mu)^2$ through an angle 2π in an anticlockwise sense

$$ A^{(+)} \to A^{(-)} \quad \text{by hermitian analyticity,} $$

$$ A^{(-)} \to A^{(\times)} \quad (A^{(\times)} \text{ is some new quantity),} $$

$$ \sqrt{\{s - (2\mu)^2\}} \to -\sqrt{\{s - (2\mu)^2\}}, $$

so that we have, on rearranging,

$$ A^{(\times)} - A^{(-)} = \frac{i \sqrt{\{s - (2\mu)^2\}}}{(8\pi)^2 \sqrt{s}} \int d\Gamma \, A^{(-)} A^{(\times)}, $$

or, returning to bubble notation,

$$ \text{—}(\times)\text{—} \;-\; \text{—}(-)\text{—} \;=\; \text{—}(-)\text{—}(\times)\text{—}. \qquad (4.7.3) $$

Subtracting from two particle unitarity (4.7.1) we find

$$\left(⚊⚊ - ⚊⊖⚊ \right) \left(⚊⊕⚊ - ⚊⊗⚊ \right) = 0 .$$

'Postmultiplying' by   ⚊⚊   +   ⚊⊕⚊   and using (4.7.1) we have

$$⚊⊕⚊ = ⚊⊗⚊ .$$

Thus if ⚊⊕⚊ is continued around $s = (2\mu)^2$ through an angle 4π in an anticlockwise sense it reproduces itself. So the two particle singularity is two sheeted (Oehme, 1961; Blankenbecler, Goldberger, MacDowell & Treiman, 1961).

The result can be extended to any two-particle threshold in any variable in any amplitude. The treatment of higher thresholds is more difficult because there are singularities touching the integration region of the discontinuity integral, e.g. in a three-particle integral (see (4.7.7) below) the integration range for the internal subenergies σ is

$$(2\mu)^2 \leqslant \sigma \leqslant (\sqrt{(s)}-\mu)^2.$$

Hence as s is continued around $(3\mu)^2$ the σ contour rotates through a similar angle about $(2\mu)^2$ thereby moving on to a new sheet with respect to the $\sigma = (2\mu)^2$ normal cut. This feature seems to preclude any simple behaviour of the higher thresholds, which are all thought to have an infinite number of sheets.

The K-matrix

The two-particle K-matrix is defined by the equation (Zimmermann, 1961)

$$⚊⊕⚊ - ⚊Ⓚ⚊ = \tfrac{1}{2} ⚊⊕⚊Ⓚ⚊ . \qquad (4.7.4a)$$

Because of the Fredholm nature of the kernel $\tfrac{1}{2}$ ⚊⊕⚊ this also equals

$$\tfrac{1}{2} ⚊Ⓚ⚊⊕⚊ . \qquad (4.7.4b)$$

The K-matrix automatically satisfies another equation relating it to the $(-)$-amplitude. Premultiplying (4.7.4a) by ⚊⊖⚊ and using (4.7.1) and (4.7.4a) to simplify we find

$$⚊Ⓚ⚊ - ⚊⊖⚊ = \tfrac{1}{2} ⚊⊖⚊Ⓚ⚊ , \qquad (4.7.5a)$$

$$= \tfrac{1}{2} ⚊Ⓚ⚊⊖⚊ , \qquad (4.7.5b)$$

also. If we continue (4.7.4a) around $s = (2\mu)^2$ so that K continues into K', we find, following the argument leading from (4.7.2) to (4.7.3)

$$\overline{(K)\overline{}} - \overline{(-)\overline{}} = \tfrac{1}{2}\overline{(-)\overline{}(K')\overline{}}.$$

Thus, comparing with (4.7.5a), K and K' satisfy the same equation. Repeating the argument following (4.7.3) we find that K and K' are actually equal. This means that the K-matrix is analytic at $s = (2\mu)^2$.

Gunson (1965) and Branson (1965) have pointed out that this result cannot be generalised to three-particle thresholds in a straightforward way, though Branson (1964, 1965) has suggested an alternative method in which some of the intermediate particles in the defining equation are off the mass shell.

Normal thresholds in subenergies

We now suppose that the two-particle discontinuity in a subenergy is given by[†]

$$\overline{(+)\overline{}} - \overline{(-)\overline{}} = \overline{(+)\overline{}(-)\overline{}}. \qquad (4.7.6a)$$

Here the labels ($\pm$) on the amplitude $\overline{\bigcirc\overline{}}$ refer only to the subenergy in which the discontinuity is being taken. The labels on the other variables in this amplitude must be the same in both terms on the left of this equation; in (4.7.6a) we suppose them all to be ($+$) but by analytically continuing the equation as a whole we may arrive at other combinations. The corresponding labels for the amplitude occurring in the term on the right will agree with those on the left in the case of variables for which no integration is implied in the term. Thus the left-hand subenergies will agree, but the two subenergies on the right which are not explicitly indicated are integration variables and their label is determined by a more complicated prescription mentioned later (§ 4.11).

Postmultiplying (4.7.6a) by $\overline{(+)\overline{}}$ we obtain

$$\overline{(+)\overline{}(+)\overline{}} - \overline{(-)\overline{}(+)\overline{}} = \overline{(+)\overline{}(-)\overline{}(+)\overline{}}.$$

Using (4.7.1) to simplify the right-hand side we find that it becomes

$$\overline{(+)\overline{}(+)\overline{}} - \overline{(+)\overline{}(-)\overline{}}.$$

[†] This was suggested by many authors, e.g. Ball, Frazer & Nauenberg (1962).

On cancellation we see that an alternative to $(4.7.6a)$ is

$$\text{[diagram]} \quad (4.7.6b)$$

Thus it is possible to interchange the labels $(+)$ and $(-)$ on the integral in $(4.7.6a)$ (just as in $(4.7.1)$). Similar results hold for the two particle discontinuity in any variable or amplitude (Olive, 1965). We stress that we are neglecting singularities other than normal thresholds in writing these equations.

Three-particle thresholds†

Equation $(4.6.6)$ for the three-particle discontinuity can be written in the form

$$\text{[diagram]} \quad (4.7.7)$$

In $\equiv\!\bigcirc\!\equiv$ the left and right entries refer respectively to the total energy and to the right subenergies taken collectively, with the labels $(+)$, $(-)$ and (i) defined by Fig. 4.7.2. In $\equiv\!\bigcirc\!=$ the left entry refers to the left subenergies and the remaining entry to the total energy.

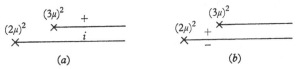

Fig. 4.7.2. (a) Boundary values in the total-energy variable. (b) Boundary values in the subenergy variables.

We shall suppose that the same discontinuity in $\equiv\!\bigcirc\!=$ is given by

$$\text{[diagram]} \quad (4.7.8)$$

Reading from left to right the three entries in each bubble refer to left subenergies, the total energy and the right subenergies respectively. Later we shall find, by adding formulae like $(4.7.6)$, that

$$\text{[diagram]}$$

---

† The remainder of this section is intended to give further insight into the structure of the theory and may be omitted at a first reading.

Using this to substitute for the right-hand bubble in (4.7.8) we find that the right-hand side of that equation becomes

$$\text{(diagram)} \quad + \quad \Sigma\ \text{(diagram)}.$$

Using another formula obtained similarly and continued around $s = (3\mu)^2$

$$\text{(diagram)} \quad - \quad \text{(diagram)} \quad = \quad \Sigma\ \text{(diagram)}$$

we find

$$\text{(diagram)} \quad - \quad \text{(diagram)} \quad = \quad \text{(diagram)}. \tag{4.7.9}$$

Comparing (4.7.8) and (4.7.9) we see that the internal subenergy boundary values can be interchanged. Continuing (4.7.8) around $(2\mu)^2$ in each right subenergy we find

$$\text{(diagram)} \quad - \quad \text{(diagram)} \quad = \quad \text{(diagram)}. \tag{4.7.10}$$

This has the form of a Fredholm integral equation (Smithies, 1958) $H - K = HK$. According to the supposed analytic properties the kernel is square integrable and hence $KH = HK$, that is

$$\text{(diagram)} \quad - \quad \text{(diagram)} \quad = \quad \text{(diagram)}. \tag{4.7.11}$$

Comparing (4.7.10) and (4.7.11) we see it is possible to interchange the label specifying the energy boundary values also. Thus there are four different forms for the three-particle discontinuity in (4.7.8), arising from the possibility of interchanging energy or internal subenergy boundary values independently. Similar results can be derived for any variable or amplitude by the methods already illustrated.

Equation (4.7.6) was originally considered valid in the region of the discontinuity variable σ: $(2\mu)^2 \leqslant \sigma < (3\mu)^2$, but it can be continued to higher values of σ following an $(i\epsilon)$-prescription in σ. It then gives the discontinuity across the $(2\mu)^2$ cut arranged as in Fig. 4.7.1.

Similarly, equation (4.7.7) for the $(3\mu)^2$ discontinuity was originally considered valid between the three- and four-particle thresholds in the total energy s. The continuation to higher values of s is slightly more complicated than in the two-particle case because the internal subenergies, which are integration variables, have as an upper end-point $(\sqrt{s}-\mu)^2$. This coincides with the singularity $(n\mu)^2$ whenever

$$s = ((n+1)\mu)^2.$$

Since $(\sqrt{(s+i\epsilon)}-\mu)^2 = (\sqrt{s}-\mu)^2+i\epsilon'$, the s-distortion determines the end-point distortion to be of like kind. Thus it is essential to arrange all normal cuts in the plane of each variable in a similar way.

More complicated singularities than normal thresholds can also be generated in the continuations. Just as in the above case the way the integral as a whole is continued around the singularity determines the way in which the end-points in the integration variable avoids the singularities of the integrand. This cannot be considered before the more complicated singularities themselves are discussed, and the matter will not be dealt with here (but see § 4.11).

Addition of discontinuities

It is thought that hermitian analyticity relates the physical $A^{(+)}$-amplitude to the $A^{(-)}$-amplitude by a path encircling all the normal thresholds corresponding to all open channels in all variables. The unitarity equations (4.2.25) give the total discontinuity, $A^{(+)}-A^{(-)}$, across all the normal threshold cuts as a sum of integrals. We now want to see how the formulae we have considered add up to give this total discontinuity. In doing so we shall try to construct integrals with (+) and (−) labels on the left- and right-hand bubbles respectively.

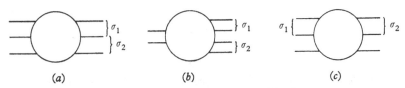

Fig. 4.7.3. Definitions of variables used in the text.

We shall deduce expressions for the compound discontinuities across cuts, in two different variables, and verify that the result is independent of the order in which the singularities are encircled to get the new boundary values. This confirms our earlier conclusion that paths encircling normal thresholds in different variables commute.

The representative examples we consider are shown in Fig. 4.7.3.

We shall specify only the boundary values in the two subenergies of interest, σ_1 and σ_2 in Fig. 4.7.3a, using the notation of Fig. 4.7.1 for the labels. According to (4.7.6) the discontinuities† in σ_1 and σ_2 across the $(2\mu)^2$ cut are

$$\qquad\qquad\qquad\qquad\qquad\qquad\qquad\qquad\qquad (4.7.12)$$

$$\qquad\qquad\qquad\qquad\qquad\qquad\qquad\qquad\qquad (4.7.13)$$

Continuing (4.7.12) around $\sigma_2 = (2\mu)^2$ we find

$$\qquad\qquad\qquad\qquad\qquad\qquad\qquad\qquad\qquad (4.7.14)$$

Adding (4.7.13) and (4.7.14) gives the desired compound discontinuity

$$\qquad\qquad\qquad\qquad\qquad\qquad\qquad\qquad\qquad (4.7.15)$$

The symmetry of this result in σ_1 and σ_2 shows that we would have obtained the same result if we had interchanged their roles in the argument, and encircled their normal thresholds in the reverse order.

The argument above is particularly simple because the variables σ_1 and σ_2 'overlap', in the sense that according to their definition (4.3.16) they involve a common momentum. Consider now the non-overlapping variable σ_1 and σ_2 of Fig. 4.7.3b. In a similar notation the $(2\mu)^2$ discontinuities are

$$\qquad\qquad\qquad\qquad\qquad\qquad\qquad\qquad\qquad (4.7.16)$$

$$\qquad\qquad\qquad\qquad\qquad\qquad\qquad\qquad\qquad (4.7.17)$$

In this case the σ_2 variable has nothing to do with the integration and if we continue (4.7.16) around $\sigma_2 = (2\mu)^2$ we see that the σ_2 boundary-value on the right-hand side is affected:

$$\qquad\qquad\qquad\qquad\qquad\qquad\qquad\qquad\qquad (4.7.18)$$

Adding (4.7.17) and (4.7.18) we find the result

$$\qquad\qquad\qquad\qquad\qquad\qquad\qquad\qquad\qquad (4.7.19)$$

---

† We have left blank the label for the σ_2 internal subenergy on the right-hand side of (4.7.8) and (4.7.9) for reasons stated earlier. A discussion of the correct assignment has been given by Hwa (1964). See also Landshoff, Olive & Polkinghorne (1966a).

which is not manifestly symmetrical in σ_1 and σ_2. But if we substitute

for the ⟨diagram⟩ amplitude in the last term, by (4.7.17), we have

$$\text{(diagram)} - \text{(diagram)} = \text{(diagram)} + \text{(diagram)} - \text{(diagram)},$$

which does have the desired symmetry. (4.7.20)

Similar arguments give (see Olive, 1965)

$$\text{(diagram)} - \text{(diagram)} = \text{(diagram)} + \text{(diagram)}. \quad (4.7.21)$$

Such analyses can obviously be extended. The addition of discontinuities across singularities in the same variable can be treated by reversing the argument given at the beginning of §4.6.

Three particle unitarity, (4.2.25b), states

$$\text{(diagram)} - \text{(diagram)} = \text{(diagram)} + \text{(diagram)} + \Sigma\,\text{(diagram)} + \Sigma\,\text{(diagram)}$$
$$+ \Sigma\,\text{(diagram)}.$$

Terms resembling the right-hand side of (4.7.15) are included under the sum in the third term on the right-hand side. Similarly, the third and fourth terms of the equation resemble the right-hand side of (4.7.21). The unitarity equation (4.2.25c) for ⟨diagram⟩ has appearing on its right-hand side the following terms (amongst others)

$$\text{(diagram)} + \text{(diagram)} + \text{(diagram)}$$

and these resemble the right-hand side of (4.7.20).

Each time, the terms mentioned are the only ones on the right-hand side of the unitarity equations involving just the relevant intermediate states. The boundary values are similar but not quite the same. As more normal threshold discontinuities are added we can expect to reproduce progressively more terms of the unitarity equation together with a better agreement in terms of boundary values until the total discontinuity is found.

The conclusion is that the extremely complicated-looking unitarity equations can be understood as a sum of the relatively simple formulae for the discontinuities across the individual normal thresholds.

4.8 Antiparticles, crossing and the TCP theorem

We saw, in §4.5, that there is a one-one correspondence between stable particles and physical-region poles of multiparticle amplitudes. We now show that the presence in one part of the physical region of a pole corresponding to a stable particle P demands the presence of an accompanying pole in another part of the physical region. From the results of §4.5, this pole must also correspond to a stable particle. This particle has the same mass as P, but the opposite intrinsic quantum numbers; it is the *antiparticle* of P. The argument, which is based on an idea of Gunson (1965), was developed by Olive (1964).

According to the results of §4.5, the four-particle→four-particle amplitude possesses a physical-region pole,

$$\begin{matrix}\text{(diagram)}\end{matrix} \qquad\qquad (4.8.1)$$

that is a pole at $(p_1+p_2-p_5-p_6)^2 \equiv s_{1256} = \mu^2$. This pole is interpreted as corresponding to the two successive interactions

$$P_1+P_2 \to P_5+P_6+P, \qquad\qquad (4.8.2a)$$

$$P+P_3+P_4 \to P_7+P_8, \qquad\qquad (4.8.2b)$$

where P_i denotes the external particle whose momentum is p_i and P denotes the internal particle, whose momentum is $(p_1+p_2-p_5-p_6)$.

Consider now the physical region for the amplitude ▭. From equation (4.3.8) the variables s_{1256}, $s_{12} = (p_1+p_2)^2$ and $s_{56} = (p_5+p_6)^2$ must satisfy, in the physical region,

$$s_{1256} \leqslant (\sqrt{s_{12}} - \sqrt{s_{56}})^2, \qquad\qquad (4.8.3a)$$

$$s_{56} \geqslant (\mu_5+\mu_6)^2, \qquad s_{12} \geqslant (\mu_1+\mu_2)^2. \qquad\qquad (4.8.3b)$$

If we fix s_{56} at a value greater than $(\mu_5+\mu_6)^2$ we may draw, in the real (s_{1256}, s_{12}) plane, the part of the boundary of the physical region derived from the constraints (4.8.3). When the fixed value of s_{56} is also larger than $(\mu_1+\mu_2+\mu)^2$ the part of the boundary in question is the heavy line in Fig. 4.8.1.† Inequalities similar to (4.8.3) are also applicable to the variables s_{1256}, $s_{78} = (p_7+p_8)^2$ and $s_{34} = (p_3+p_4)^2$ and can be represented similarly.

† Compare with (4.3.20) and Fig. 4.3.1.

In the physical regions for the processes (4.8.2*a*) and (4.8.2*b*) we must have, respectively

$$\sqrt{s_{12}} \geqslant \sqrt{(s_{56})} + \mu, \tag{4.8.4a}$$

$$\sqrt{s_{78}} \geqslant \sqrt{(s_{34})} + \mu. \tag{4.8.4b}$$

This corresponds to the part of L (the line $s_{1256} = \mu^2$ in Fig. 4.8.1) above A. So only on this part of L do we get a pole in the physical region of that may be interpreted as being due to the successive interactions (4.8.2).

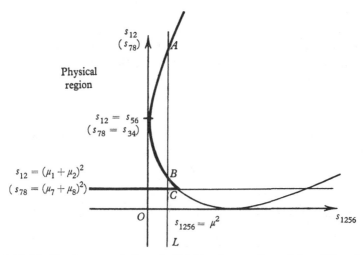

Fig. 4.8.1. The line L represents the pole (4.8.1) and passes through two different parts of the physical region for the four-particle → four-particle amplitude. The boundary of the physical region is drawn in heavy line.

Yet L enters the physical region of a second time, on the segment BC, when

$$\sqrt{s_{12}} \leqslant \sqrt{(s_{56})} - \mu, \tag{4.8.5a}$$

$$\sqrt{s_{78}} \leqslant \sqrt{(s_{34})} - \mu, \tag{4.8.5b}$$

and since poles persist in functions of several complex variables this will give another physical region pole, provided that in continuing along L we have not been forced on to the wrong side of a branch-point so that the pole is on a Riemann sheet different from the physical sheet. If this proviso is satisfied the pole in this second part of the physical region must correspond to a particle $\bar{P}$ which also has mass μ.

However, because of the inequalities (4.8.5) it now corresponds to two successive interactions

$$P_3 + P_4 \rightarrow P_7 + P_8 + \bar{P}, \\ P_1 + P_2 + \bar{P} \rightarrow P_5 + P_6, \} \qquad (4.8.6)$$

rather than (4.8.2). Diagrammatically we have

$$(4.8.7)$$

The particle $\bar{P}$, although it has the same mass μ as P is generally different from P. As may be seen by comparison of (4.8.2) and (4.8.6) the additive quantum numbers of P and $\bar{P}$ (charge, strangeness, etc.) must be opposite. Hence we identity $\bar{P}$ as the antiparticle of P and conclude that antiparticles must exist in the theory.

This establishes the existence of antiparticles except for the crucial question as to whether L is actually a pole in the part of the physical region satisfying (4.8.5). To prove this we want a path of continuation joining the two parts of the physical region but lying always on the section $s_{1256} = \mu^2$. Then we can use this to analytically continue the function

$$[(s_{1256} - \mu^2) A(s_{1256}, s_{12}, \ldots)], \quad \text{where} \quad A = \text{image}.$$

This is the residue of the pole (4.8.1) and if it is non-zero in (4.8.4) it cannot be identically zero in (4.8.5).

Because the physical region is connected, our analyticity postulate implies that there exists a path of analytic continuation joining the two regions (4.8.4) and (4.8.5) of L and lying in the physical region except for the infinitesimal detours around singularities. If this path can be deformed to lie on the mass shell section $s_{1256} = \mu^2$ without crossing any other singularity surfaces it will provide the desired continuation of the residue. In terms of a singularity structure involving just normal thresholds this is possible because normal thresholds either intersect L or are parallel to it. The physical region path Γ cannot encircle any of the parallel normal thresholds ($s_{1256} = (n\mu)^2$) because of our $+i\epsilon$-prescription assumption. Hence it can be pulled free of them, and can be slid along any of the others. However it is not yet possible to make more general arguments because the theory does not yet give definite information about the singularities in the unphysical region through which Γ must be deformed. There is reason to be optimistic nevertheless, because if the singularities in the unphysical region have properties like those in the physical region the deformation will be possible, since in the physical region the postulate that the amplitude is the boundary value of an analytic function guarantees that any path is freely deformable.

Crossing

Having deduced the existence of antiparticles, we must insert them in the completeness relations and so they will appear in the unitarity equations where their quantum numbers are appropriate. Hence transition amplitudes involving antiparticles play an important part in the theory. From our derivation of the existence of antiparticles $\bar{P}$ we see that these amplitudes are directly related to the corresponding 'crossed' amplitudes involving the particles P. Since (4.8.7) is obtained by analytic continuation of (4.8.1), the product of the two amplitudes in (4.8.2) continues analytically into the product of the crossed amplitudes in (4.8.6). We obtained this result by considering a particular multiparticle amplitude ⊒◯⊒, but we would obtain corresponding results for any multiparticle amplitude containing a pole corresponding to the particle P. Thus we conclude that, for any amplitude A having P as one of the external particles,

$$\begin{aligned} A^{\text{continued}} &= \alpha\bar{A} \quad \text{if } P \text{ is incoming for } A, \\ &= \alpha^{-1}\bar{A} \quad \text{if } P \text{ is outgoing for } A, \end{aligned} \right\} \tag{4.8.8}$$

where $\bar{A}$ is the corresponding crossed amplitude having $\bar{P}$ as an external particle, and α is a constant number whose value depends solely on the identity of the particle P. [α must be a constant, independent of the variables s, since we may change the values of the variables for one of the amplitudes in the product of amplitudes that forms the residue of the pole, without changing them in the other. It is independent of the particular amplitude, because any given amplitude involving P in the initial state may be paired with any other involving P in the final state, to form the residue of a P-pole in some multiparticle amplitude.]

We next show that α is just a phase factor, that is $|\alpha| = 1$. For definiteness, let A be the amplitude for (4.8.2a)

$$A: \quad P_1 + P_2 \rightarrow P_5 + P_6 + P,$$

so that $\bar{A}$ is the second amplitude in (4.8.6)

$$\bar{A}: \quad P_1+P_2+\bar{P}\to P_5+P_6,$$

and, by (4.8.8), $A^{\text{cont.}} = \alpha^{-1}\bar{A}.$ (4.8.9a)

Define two further amplitudes

$$A' = P_5+P_6+P\to P_1+P_2,$$

$$\bar{A}' = P_5+P_6\to P_1+P_2+\bar{P}.$$

Then, again by (4.8.8), $A'^{\text{cont.}} = \alpha\bar{A}'.$ (4.8.9b)

We have said, in §1.2, that sometimes S-matrix elements obey the symmetry property (1.2.23), so that $A = A'$ and $\bar{A} = \bar{A}'$. In this case (4.8.9a) and (4.8.9b) together give $\alpha^2 = 1$. More generally, A' is related to A, and $\bar{A}'$ to $\bar{A}$, by hermitian analyticity. According to §4.6, this property says that A'^* and $\bar{A}'^*$ are respectively obtained from A and $\bar{A}$ by analytic continuation. Hence we see that in two ways we may obtain $\bar{A}'^*$ by continuation from A:

$$A \xrightarrow[\text{crossing}]{} \alpha^{-1}\bar{A} \xrightarrow[\text{hermitian analyticity}]{} \alpha^{-1}\bar{A}'^*,$$

$$A \xrightarrow[\text{hermitian analyticity}]{} A'^* \xrightarrow[\text{crossing}]{} \alpha^*\bar{A}'^*.$$

One might expect that if two different continuations of the same function lead to functions that are proportional to one another, these continuations are actually the same, so that $\alpha^{-1} = \alpha^*$. However, this is not always true, so that it is necessary explicitly to verify that the continuations are equivalent, that is that the two paths of continuation may be continuously distorted into each other without crossing any branch-point. Again this may be shown, as far as the normal thresholds are concerned.

Having decided that $|\alpha| = 1$, we may now actually *define* $\alpha = 1$. This is because, in our initial construction in §4.2 of the momentum states by means of creation operators, there was an arbitrariness of phase. This phase could be defined independently for particles and antiparticles, and chosing $\alpha = 1$ merely links the two definitions. In the case when a particle is its own antiparticle or it belongs to an isotopic multiplet for whose members the relative phases are fixed by other conventions, additional consideration must be given to the matter. This has been discussed by J. R. Taylor (1966).

Thus we have the fundamental theorem known as *crossing*, that an amplitude for a process obtained by 'crossing over' a particle into its antiparticle is analytically related to the amplitude for the original process. Further, the path of continuation obeys a $(+i\epsilon)$-prescription at each normal threshold traversed, since the physical region path in the large amplitude had this property and it is preserved in the distortion to the mass shell section.

The TCP Theorem

So far we have only considered the effects of crossing one particle. However, for the two-particle → two-particle amplitude this would lead to an unphysical amplitude, and we must simultaneously cross two particles. The discussion is exactly parallel to the above, but now involves a twofold pole of the five-particle amplitude:

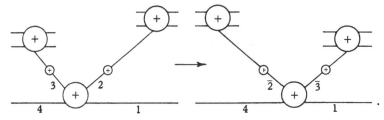

Hence it is found that the amplitude for

$$P_1 + P_2 \rightarrow P_3 + P_4 \tag{4.8.10}$$

may be continued analytically into that for

$$P_1 + \bar{P}_3 \rightarrow \bar{P}_2 + P_4, \tag{4.8.11}$$

the path of continuation following a $(+i\epsilon)$-prescription for all normal thresholds traversed. In terms of the familiar s, t and u variables defined by (1.2.10), the path in the s-plane (with t fixed) is shown in Fig. 4.8.2. Note that in order to maintain the $(+i\epsilon)$-prescription at the u normal thresholds the path in the s-plane must go under them because (1.2.12), relating u to s with t fixed (and real), implies that $\text{Im}\, s = -\text{Im}\, u$.

Similarly, the amplitude for (4.8.11) may be continued analytically into that for

$$\bar{P}_4 + \bar{P}_3 \rightarrow \bar{P}_2 + \bar{P}_1. \tag{4.8.12}$$

Hence the amplitudes for the processes (4.8.10) and (4.8.12) are analytically related by a path obeying a $(+i\epsilon)$-prescription at each normal threshold.

Since particles and antiparticles have identical masses we can choose the four-momenta of the corresponding particles in the processes (4.8.10) and (4.8.12) to be equal so that the variables s, t, u defined in the usual way take the same values for each process and the path of continuation returns to its starting point. Since this path obeys the same $(+i\epsilon)$-prescription each time it encounters a normal threshold it cannot encircle such a threshold. It can therefore be contracted to zero if these are the only singularities it encounters. This is particularly clear from Fig. 4.8.2 since the path relating the amplitude for the process (4.8.11) to that for (4.8.12) is exactly the reverse of the path drawn relating the amplitudes for processes (4.8.10) and (4.8.11). Thus the amplitudes for the two processes (4.8.10) and (4.8.12) with momenta related as described are equal. This is the *TCP theorem*. Once again the complete proof requires a better knowledge of the singularity structure.

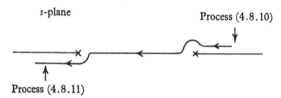

Fig. 4.8.2. The path of analytic continuation that gives crossing for a two-particle → two-particle amplitude.

The particular importance of the TCP theorem is that its proof is one of the triumphs of axiomatic field theory. (See, for example, the book by Streater & Wightman (1964)). If the present theory is to be worthwhile it must at least accomplish what field theory has achieved.

Complete normal threshold structure

In proving crossing we prove that amplitudes for certain processes are continuations of each other. It follows that singularities of one amplitude are singularities of the other and vice-versa. In this way we learn of new singularities, for example, the left-hand cut of the two particle → two particle amplitude, drawn in Fig. 1.3.4; diagrammatically this rises from

$$(4.8.13)$$

Likewise the existence of normal thresholds in those momentum transfer variables of the uncrossed amplitude which cross into subenergies follows from the unitarity and extended unitarity equations for the crossed processes:

$$(4.8.14)$$

Similarly, physical-region poles (§ 4.5) imply the existence of unphysical-region poles in crossed channels

$$(4.8.15)$$

where the factors in the residue on the right-hand side are continuations of ⊐(+)⊏ into an unphysical region. This argument, however, only applies to the multiparticle amplitudes and not to the two particle→two particle amplitude, because that amplitude has no stable particle poles in the physical region for any of its channels. By considering the system of unitarity equations (4.2.25*b*) coupling ⊐(+)⊏ to ⊐(+)⊏ it is possible to show that, because of (4.8.15), ⊐(+)⊏ does have a pole on the 'physical sheet'; this pole† is represented by the point P in Fig. 1.3.4 (with P' the corresponding pole in the crossed channel). Furthermore, the residue of this pole factorises (as it did in (4.8.15)) and we write diagrammatically

$$\text{⊐(+)⊏} \quad \sim \quad \text{⊐O—O—O⊏} , \qquad (4.8.16a)$$

corresponding to

$$A^{(+)}_{22}(s,t) \sim \frac{gg^*}{s - m^2}, \qquad (4.8.16b)$$

where A_{22} represents the two particle scattering amplitude and g is the coupling constant ⊐O—.

Adding these new results to those of the previous sections we find that we have deduced the existence of all the normal threshold and pole singularities corresponding to stable particles. The derivations of the fundamental theorems, hermitian analyticity, crossing, etc., can be repeated in the presence of these singularities.

Crossing and unitarity

The normal thresholds just mentioned occur in a similar fashion in each of the variables (4.3.16) so that the situation appears to be highly

---

† Notice that here we are discussing a particle of spin zero (see p. 253).

crossing symmetric. Nevertheless, we must check that the associated scheme of discontinuities for the normal thresholds is consistent with crossing (Olive, 1965).

The two-particle discontinuity in the subenergy s_{12} of the amplitude $\substack{4\\5\\6}$⊃(+)⊂$\substack{1\\2\\3}$ has the form (4.7.6a):

$$ \exists(+)\exists \;-\; \exists(-)\exists \;=\; \exists(+)(-)\exists. $$

Continuing this equation along the path of continuation corresponding to crossing over line 6 we find

$$ \exists(+)\exists \;-\; \exists(+)\exists \;=\; \exists(+)(-)\exists. $$

This is just like (4.7.16), another of the normal threshold discontinuity formulae, and illustrates the general fact that the formulae for normal threshold discontinuities that we discussed in §4.7 are 'crossing symmetric' in the sense that they cross into each other.

Consider two points A and B in the physical regions of two multi-particle processes which can be crossed into each other. Of the channels open in the various variables at the two points A and B there will be a certain common subset. The total discontinuity across the corresponding normal thresholds we shall call the 'common discontinuity'. We saw in §4.7 that the expressions for this common discontinuity in the two amplitudes will resemble those terms in the corresponding unitarity relations which involve only the common intermediate states, but will differ in the assignment of boundary values.

Since, as we saw in §4.7, the common discontinuity is the sum of the individual discontinuities across the cuts attached to the common thresholds and since we have just seen that these discontinuities are themselves 'crossing symmetric', so must be the common discontinuity itself. In this sense we can say the unitarity equations are 'crossing symmetric'.

As an illustration consider the amplitudes

$$ \substack{4\\5\\6}⊃(+)⊂\substack{1\\2\\3}, \qquad \substack{\bar3\\6}⊃(+)⊂\substack{\bar4\,5\\1\,2}, $$

which can be related by crossing over lines 3, 4 and 5. The relevant unitarity relations valid just above the physical threshold appear in (4.2.25), and the common normal thresholds are $s_{12} = (2\mu)^2$ and

$s_{45} = (2\mu)^2$. According to (4.7.21) and (4.7.20) the respective common discontinuities are

$$(4.8.17)$$

$$(4.8.18)$$

Continuing (4.8.17) along the path corresponding to crossing lines 3, 4 and 5 we obtain not (4.8.18) but

Nevertheless on applying the boundary value interchange (4.7.6 a, b) to the second term on the right-hand side we obtain (4.7.19), which reduces to (4.7.20), that is (4.8.18), on application of the discontinuity formula (4.7.17), as we have already seen.

Thus, though at first sight the crossing symmetric property of the unitarity equations looks unlikely, it is nevertheless true.

4.9   Unstable particles

Since the asymptotic states on which our theory is based are required to be states that exist for an arbitrarily long time, they can contain only stable particles. Hence only stable particles enter directly into the completeness relation and the unitarity equations. Apart from these considerations, however, we shall see in this section that, mathematically, the theory involves the unstable particles in a very similar way to the stable ones. This is to be expected, from the fact that it is very difficult to distinguish physically between an unstable particle of long lifetime and a stable particle.

Since a basic property of an unstable particle is that it has a lifetime, a full discussion of its properties must contain some mention of this. However, as we explained in § 4.5, the role of time in S-matrix theory still requires considerable clarification, so here we only give a very crude and very brief discussion. For a plane-wave state of energy E the dependence on the time τ of the Schrödinger wave-function is given by the factor $e^{-iE\tau}$. In the rest frame of the particle this is $e^{-iM\tau}$. For an unstable particle the wave-function should decrease with increasing time. This is most simply achieved by making M complex, giving it a negative imaginary part which will then be proportional to the inverse of the lifetime. As we shall see below, the assignment of a complex mass to an unstable particle is also required by other considerations.

Another basic property of an unstable particle of short lifetime, and the one by means of which it is usually observed, is that it corresponds to a *resonance* in interactions among the stable particles. For example, if two stable particles A scatter at centre-of-mass energy $\sqrt{s}$ near to the 'mass' of an unstable state with appropriate quantum numbers, they can then form that state and remain in it for a time, roughly equal to its lifetime, before it decays again. Hence the incoming wave-function is greatly distorted and the transition amplitude for the scattering $A + A \rightarrow A + A$ is enhanced, that is it has a bump at the appropriate energy. The simplest way an analytic function can have a bump for values of s on the real axis is for there to be an infinite singularity nearby, off the real axis. The simplest such singularity would be a simple pole, that is a pole at the complex point $s = M^2$ in the lower-half plane. The bump in the amplitude then occurs around $s = \mathrm{Re}\,(M^2)$ or, what is much the same thing if the lifetime is not too short and so $\mathrm{Im}\,M$ not too large, around $s = (\mathrm{Re}\,M)^2$. So we identify $(\mathrm{Re}\,M)$ as the physical mass of the unstable particle, the vagueness in the identification corresponding directly to the fact that the uncertainty principle does not allow a particle having only a finite lifetime to have a definite mass.

That unstable particles really are represented by complex poles was suggested by Møller (1946), and is shown to be the case in the analysis below. The corresponding bumps in cross-sections are then given by the Breit–Wigner resonance formula (1.6.1).

Variable-force model

The simplest way to discover what should be the properties of the singularities corresponding to unstable particles is to make use of a model (Landshoff, 1963). We shall show later in this section that the simpler predictions of the model can be checked by more rigorous methods.

It is supposed that the theory contains a parameter λ that measures the strength of the forces between the stable particles A. It is further supposed that when λ is varied the theory changes analytically, and that it can be varied in such a way that the forces become so strong that the unstable particle becomes stable.

Under these reasonable, though unsubstantiated, assumptions it follows immediately that the unstable particle results in a pole. This is because we know that it corresponds to a pole when the particle is stable (see §4.5 and also (4.8.16)), and therefore likewise when it is unstable since the nature of singularities cannot change† during the analytic continuation with respect to λ. Further, when the particle is stable we know that it gives rise to other singularities, the normal thresholds and the more complicated singularities to be discussed in

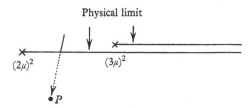

Fig. 4.9.1. The unstable-particle pole P is reached from the physical region by going down through the cut.

§4.10, and so in the same way it must generate similar singularities when it is unstable. The corresponding discontinuities in the unstable case will be obtained from those in the stable case by analytic continuation.

To determine on what Riemann sheets these singularities lie we must consider what path the pole P takes when the particle that it represents changes from being stable to being unstable. We have said that, in the unstable case, the pole is in the lower-half of the complex s-plane and, in order that it may produce a bump in the amplitude, that it is near to the physical region. Hence it must be on the unphysical Riemann sheet that is reached from the physical region by burrowing down through the normal-threshold cut. This we have indicated in Fig. 4.9.1, where we have supposed that the 'mass' of the unstable particle is below the $(3\mu)^2$-threshold. Otherwise, to reach the pole from the physical region, it would be necessary to pass through both the $(2\mu)^2$ and $(3\mu)^2$ cuts and the subsequent analysis would require corresponding changes.

† Except possibly for discrete values of λ; it would be an interesting and highly powerful principle if physics were to choose just those values to represent reality.

Hence, since we know that in the stable case the pole P at $s = M^2$ is on the physical sheet, in the transition to instability it must pass round the $(2\mu)^2$ branch-point and through the cut, as indicated in Fig. 4.9.2a. In this figure the part of the path on the physical sheet is drawn with a dashed line, the part on the unphysical sheet with a dotted line. Having decided that the point $s = M^2$ takes this path, we see directly that the normal-threshold $s = (M + \mu)^2$, whose existence

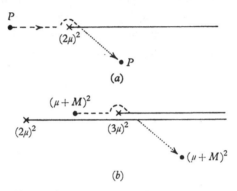

Fig. 4.9.2. (a) A possible path taken by the pole P in the transition from stable to unstable state. (b) The corresponding path for the normal threshold.

in the stable case follows from unitarity, must take the path shown in Fig. 4.9.2b. Because $s = M^2$ goes round the point $(2\mu)^2$, $s = (M + \mu)^2$ must go round $(3\mu)^2$, and so the branch-point in the unstable case is reached from the physical region by going straight down through the cuts. As we saw in §4.7, this branch-point is of square root type and does not give an infinity in the amplitude, so it will not produce such a noticeable effect in the physical amplitude as does the pole.†

For simplicity, we have glossed over certain points in drawing Fig. 4.9.2a. Hermitian analyticity demands that, when P is complex, there be a 'shadow' pole P' in the complex-conjugate position, on the unphysical sheet reached from the physical sheet by going upwards through the $(2\mu)^2$ cut.‡ The path that P' takes in the transition, corresponding to that of Fig. 4.9.2a taken by P, is drawn in Fig. 4.9.3a; while P is complex P' must occupy the complex-conjugate position, but when P is real so is P'. In the latter case P' and P are on different sheets, because otherwise a particle of the mass corresponding to P'

† The effect is to produce a 'woolly cusp'; see Nauenberg & Pais (1962).
‡ Since, as we saw in §4.7, this cut happens to be two-sheeted, in this particular case P and P' are actually on the same sheet.

would have to appear in the completeness and unitarity relations; they are also in different positions, as we see later, in § 4.10.

Generally, our model theory could only be expected to be capable of representing a physical situation for real values of the parameter λ. This is the case when, for example, λ represents the width or the depth of a potential, or when it is a coupling constant. Continuations to complex values of λ cannot be expected to satisfy unitarity or hermitian analyticity. But if we vary λ through real values to effect the

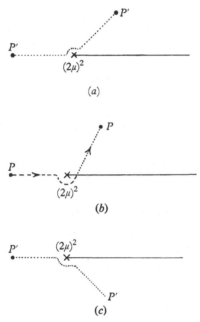

Fig. 4.9.3. (a) The path taken by the shadow pole P' when P takes that of Fig. 4.9.2 a. P' remains on the unphysical sheet. (b) An alternative path for P, obtained by giving λ a small imaginary part of opposite sign from before. (c) The corresponding path for P'.

transition from a stable to an unstable state, the pole P will pass through the branch-point, instead of round it as we have drawn in Fig. 4.9.2 a. In this case one cannot tell whether or not it has changed its Riemann sheet, so to achieve the unambiguous path of Fig. 4.9.2 a we temporarily give λ a small imaginary part (so temporarily abandoning a physically relevant configuration) to make the pole avoid the branch-point. It is interesting to notice that if we had chosen to give λ an imaginary part of the opposite sign, the pole P would have gone to the other side of the branch-point, as in Fig. 4.9.3 *b*. Similarly, the path for P', instead of being as in Fig. 4.9.3 *a*, would be as in Fig. 4.9.3 *c*, so now it is P' instead of P that represents the unstable particle.

We have deduced that every pole is accompanied by at least one shadow pole. In fact there is reason to believe that for a given pole there are an infinite number of shadow poles, on different Riemann sheets and in a different position on each sheet. [See Landshoff (1963); also Eden & Taylor (1964).] It is therefore something of a simplification to talk about *the* pole corresponding to a given particle, though generally one pole will be physically more significant than all the others, in that it lies closer to the physical region. All the shadow poles are to be expected to generate corresponding 'normal thresholds' and higher singularities.

Finally, it should be mentioned that it is not always the case that, when λ is varied, the transition is directly from unstable to stable state; it may take place via a 'virtual' state, that is the pole first meets the real axis on the unphysical sheet, before encircling the threshold and passing on to the physical sheet. This makes little difference to the deductions concerning the further singularities; more details are given by Landshoff (1963).

Complex poles from unitarity

The model that we have described is useful for discovering what results are expected to hold; to prove them we use unitarity. We first examine how the complex poles arise (Gunson & Taylor, 1960).

Consider the unitarity equation

$$\text{---(+)---} - \text{---(−)---} = \text{---(+)------(−)---} = \text{---(−)------(+)---}. \qquad (4.9.1)$$

We wish to continue this equation as a whole in the total energy variable *s*. We take a path of continuation corresponding to the arrow in Fig. 4.9.1, so that for ---(+)--- we are moving on to the unphysical sheet but for ---(−)--- are staying on the physical sheet. A proper analysis involves the theory of Fredholm integral equations, but it turns out that the properties of interest here may be understood by replacing (4.9.1) by a similar equation involving finite matrices:

$$H(s) - K(s) = H(s)\,K(s) = K(s)\,H(s). \qquad (4.9.2)$$

The matrix product on the right-hand side of this equation is the finite-dimensional analogue of the integration in the right-hand side of (4.9.1).

Solving for H in terms of K, we have

$$H = K(1-K)^{-1} = (1-K)^{-1}K$$
$$= \frac{K\operatorname{adj}(1-K)}{\det(1-K)} = \frac{\operatorname{adj}(1-K).K}{\det(1-K)}. \tag{4.9.3}$$

The assumption that $\operatorname{\underline{\hspace{0.3em}}\bigcirc\operatorname{\underline{\hspace{0.3em}}}$ is analytic on the physical sheet, except for the cuts and poles on the real axis, produces similar analytic properties for K. From (4.9.3) we see that H is similarly analytic, except for poles arising from zeros of $\det(1-K)$. Since H represents the continuation of $\operatorname{\underline{\hspace{0.3em}}\bigcirc\operatorname{\underline{\hspace{0.3em}}}$ on the unphysical sheet these are the poles of interest.

By Jacobi's identity (4.3.14), at a zero of $\det(1-K)$ the matrix $\operatorname{adj}(1-K)$ factorises. Hence by (4.9.3) the residue matrix R, defined by

$$R = \lim_{s \to M^2} [(s-M^2)H],$$

also factorises:
$$R_{ij} = a_i b_j. \tag{4.9.4}$$

This confirms what we deduced from our model, that the residues of the unstable-particle poles have properties exactly similar to those of stable particles, in particular factorisation.

The corresponding results from Fredholm theory are not quite so simple. One feature is that the matrix R, now of infinite dimension, does not necessarily factorise and so have rank one; it may have finite rank greater than one (Smithies, 1958). This rank is the number of different spin states of the unstable particle; we do not consider it here, and confine the discussion to spin zero.

Corresponding to equation (4.8.16) for stable particle poles we write

$$\operatorname{\underline{\hspace{0.3em}}}\!\oplus\!\operatorname{\underline{\hspace{0.3em}}} \;\sim\; \operatorname{\underline{\hspace{0.3em}}}\!①\!\sim\!\sim\!\bigcirc\!\sim\!\sim\!②\!\operatorname{\underline{\hspace{0.3em}}}, \tag{4.9.5a}$$

or equivalently
$$A_{22}^{(+)} \sim \frac{G_1 G_2^*}{s - M^2}, \tag{4.9.5b}$$

where G_1 and G_2 are the coupling constants coupling the unstable particle P to the two stable particles A.

This analysis leaves open two questions. It does not tell us that the complex pole must be on the unphysical sheet rather than the physical sheet (as potential theory and field theory suggest), for we *assumed* that K had no complex poles. Neither does it tell us that the pole is simple rather than multiple (see Eden & Landshoff, 1964).

The analogy between stable and unstable particle can be extended by defining amplitudes for processes involving unstable particles

(Gunson, 1965; Stapp, 1964a). Consider the discontinuity equation for the $(2\mu)^2$ threshold in the variable s_{12} of the amplitude (diagram) $\frac{1}{3}\frac{2}{4}$ (see (4.7.6)),

$$\text{(diagram)} - \text{(diagram)} = \text{(diagram)}, \qquad (4.9.6a)$$

$$= \text{(diagram)}. \qquad (4.9.6b)$$

If we continue (4.9.6b) into the lower half plane of the variable s_{12} we shall arrive at a pole of the right-hand side, because of (4.9.5). We expect that (diagram), like (diagram), will not possess this pole because s_{12} moves on to its physical sheet. Then (diagram) must have a pole at $s_{12} = M^2$, whose residue factorises because the residue of the pole on the right-hand side of (4.9.6b) factorises. In fact, near $s_{12} = M^2$ we must have

$$\text{(diagram)} \sim \text{(diagram)}, \qquad (4.9.7)$$

where

$$\text{(diagram)} = \text{(diagram)}. \qquad (4.9.8)$$

We saw in §4.5 that the factors in the residue of a stable particle pole were themselves amplitudes involving this particle. This suggests that we can treat the quantity (diagram) as an amplitude describing a process involving an unstable particle. Similarly, we can define amplitudes (diagram), (diagram), and so on.

The methods of §4.8 can be developed to prove crossing for these unstable amplitudes. Furthermore, discontinuities in these amplitudes are given by equations of the now familiar type (see §4.7). As an example consider the continuation of (4.9.6), evaluated in the $+$ boundary value in s_{34}, to the lower half plane in that variable. Each term has a pole with factorising residue and we obtain, cancelling common factors

$$\text{(diagram)} - \text{(diagram)} = \text{(diagram)}$$

$$= \text{(diagram)}. \qquad (4.9.9)$$

Complex normal thresholds

Having seen how the presence of complex poles fits in with unitarity, we now examine the corresponding normal thresholds and calculate the associated discontinuities (Gunson, 1965; Zwanziger, 1963). We already know from our model that the discontinuity should turn out to be exactly similar to that for an ordinary stable particle normal threshold.

The model predicted that the complex normal threshold occurred on the unphysical sheet reached by burrowing through the three-particle cut. Hence we must study equations for the discontinuity across this cut. The most convenient forms to consider are (4.7.10) and (4.7.11).

$$\text{=}\!\!\bigoplus_{+|+|-}\!\!\text{=} \; - \; \text{=}\!\!\bigoplus_{i|-}\!\!\text{=} \; = \; \text{=}\!\!\bigoplus_{+|i}\!\!\text{=}\!\!\bigoplus_{+|+|-}\!\!\text{=}, \qquad (4.9.9a)$$

$$\text{=}\!\!\bigoplus_{+|+|-}\!\!\text{=} \; - \; \text{=}\!\!\bigoplus_{i|-}\!\!\text{=} \; = \; \text{=}\!\!\bigoplus_{+|+|-}\!\!\text{=}\!\!\bigoplus_{i|-}\!\!\text{=}. \qquad (4.9.9b)$$

In each bubble, the centre label refers to the boundary limit taken for the total-energy variable s. As before, the label (i) denotes the boundary value obtained by continuing from the physical limit $(+)$ round to the lower side of the three-particle cut, as is shown in Fig. (4.7.2a). The other labels refer to the subenergies.

We now continue $(4.9.9a)$ into the lower half of the complex s-plane. If we label the internal lines in the term on the right-hand side by 1, 2, 3 the extremum of the integration over the internal subenergy s_{ij} is $(\sqrt{s} - \mu_k)^2$, so this also goes into the lower-half plane. (It is now convenient to suppose that the masses μ_i of the internal particles are all different.) So far as the internal subenergies s_{ij} are concerned, we are on the physical sheet in our continuation of the left-hand bubble of the integrand, since the labels $(-)$ mean that these variables started underneath their cuts and so the continuation is not taking them through the cuts. However, for the right-hand bubble these variables started on the upper side of their cuts and so the continuation takes them just on to that sheet which has the pole

$$\text{=}\!\!\bigoplus_{+|+|-}\!\!\text{=} \; \sim \; \underline{\qquad}\!\!\bigoplus_{+|-}\!\!\text{=}, \qquad (4.9.10)$$

at $s_{12} = M^2$ (cf. (4.9.7) and Fig. 4.9.1). When the extremum $(\sqrt{s} - \mu_3)^2$

of s_{12} strikes this pole the multiple integral has a singularity.† This occurs at $s = (M + \mu_3)^2$.

To calculate the discontinuity associated with this complex normal threshold, we continue (4.9.9a) to either side of the corresponding cut and compare the results. We here accept the prediction of our model, that the normal threshold occurs on the Riemann sheet of s reached by continuation straight down into the lower-half plane for [diagram: $+$] but not for [diagram: i]. This means that the two required continuations of the latter are identical where it appears on the left-hand side of (4.9.9a). They are also identical where it appears on the right because of the additional feature that the pole at $s_{12} = M^2$ that generates the singularity occurs only in the continuation of the factor [diagram: $+$ $+$ $-$] in the integral. So the two continuations of (4.9.9a) are

$$[\text{diagram: } + U] - [\text{diagram: } + i -] = [\text{diagram: } + i -][A U -], \qquad (4.9.11a)$$

$$[\text{diagram: } + L] - [\text{diagram: } + i -] = [\text{diagram: } + i -][B L -]. \qquad (4.9.11b)$$

Here U, L respectively denote the continuations in the total-energy variable s from ($+$) to the upper and lower sides of the cut, while A, B respectively note that the s_{12} integration passes above and below the pole on the s_{12} unphysical sheet. The other labels now represent straightforward continuations from their previous meanings. (In the case of the subenergies, continuation is only forced on us in the case of the internal ones, but we can choose to continue the others also.)

Subtraction of (4.9.11a) from (4.9.11b), and rearrangement, gives

$$[\text{diagram: } + U] - [\text{diagram: } + L] = [\text{diagram: } + i -] \left[\left([A U -] - [B U -] \right) + \left([B U -] - [B L -] \right) \right].$$

$$(4.9.12)$$

The first bracket within the square bracket produces the difference between integrations that pass to either side of the pole. This just gives $-2\pi i$ times the residue (4.9.10). We can write this term as

$$[\text{diagram: } + i -][U -]. \qquad (4.9.13)$$

† See §2.1 for a detailed discussion of singularities in multiple integrals.

This is to be interpreted by the rules (4.2.30) with, loosely, the line ∿∿ representing the factor $-2\pi i\delta(s_{12}-M^2)$, in close analogy with the meaning we attach to internal stable-particle lines. We say 'loosely', because M^2 is complex and so the usual meaning of the δ-function must be extended, in practice by analytic continuation of the integral from the case where M^2 is taken as real.

By an identity analogous to (4.9.8), (4.9.13) is the same as

$$\exists\!\!\!\in\!\!\boxed{+\,|\,i\,|}\!\!\curvearrowright\!\!\boxed{U\,|-}\!\!\in,$$

so, on rearranging (4.9.12), we have

$$\left(\equiv\equiv\ -\ \exists\boxed{+|i|}\in\right)\!\left(\exists\boxed{+|U|}\in\ -\ \exists\boxed{+|U|}\in\right)\ =\ \exists\boxed{+|i|}\curvearrowright\boxed{U|-}\in.$$

We premultiply this by

$$\equiv\equiv\ +\ \exists\boxed{+|L|-}\in,$$

and use the 'mate' of (4.9.11b), obtained by continuation of (4.9.9b) instead of (4.9.9a). This gives

$$\exists\boxed{+|U|}\in\ -\ \exists\boxed{+|L|-}\in\ =\ \left(\equiv\equiv\ +\ \exists\boxed{+|L|-}\in\right)\!\left(\exists\boxed{+|i|}\curvearrowright\boxed{U|-}\in\right).$$

$$(4.9.14)$$

Now, just as we derived (4.9.9) from (4.9.6), so from the 'mate' of (4.9.9b) we can immediately obtain

$$\exists\boxed{+|L|}\curvearrowright\ -\ \exists\boxed{+|i|}\curvearrowright\ =\ \exists\boxed{+|L|-}\!\!\exists\boxed{+|i|}\curvearrowright. \qquad (4.9.15)$$

Then (4.9.14) and (4.9.15) give the final result

$$\exists\boxed{+|U|-}\in\ -\ \exists\boxed{+|L|-}\in\ =\ \exists\boxed{+|L|}\curvearrowright\boxed{U|-}\in, \qquad (4.9.16)$$

which has the expected form. From this expression the corresponding discontinuities in the amplitudes $\sqsupset\!\!\bigcirc\!\!\sqsubset$ and $\sqsupset\!\!\bigcirc\!\!\sqsubset$ can also be derived.

The parallels between the properties of the unstable and stable particles suggest that they are likely to enter dynamical calculations on an equal footing. On the other hand, the ambiguity in the position of the pole associated with an unstable particle seems to make it difficult to include them in any basic axioms.

4.10   Generation of singularities

In §§ 4.6 to 4.9 we have only taken into account a singularity scheme involving the poles and normal thresholds associated with stable and unstable particles, the relevant discontinuities being determined in a crossing-symmetric way from the physical unitarity equations. Apart from the notational problem of stating the arguments completely generally rather than by means of illustrative examples, this work seems to have been taken as far as possible within the limitation of a purely normal threshold discussion.

Certain questions remain to be asked:

(a) Are there other singularities? If so, how are they generated and what are their discontinuities?

(b) Can the derivation of fundamental results like crossing, hermitian analyticity, etc., be carried through in the presence of higher singularities?

(c) Can we justify the procedure of building up the theory in a series of stages, of which the normal threshold picture is the first?

The first question was answered by Polkinghorne (1962a, b). The minimum singularity structure is determined by unitarity and crossing. The way that unitarity immediately leads to normal thresholds has already been described; however, the non-linearity of the unitarity relations generates further singularities.

In § 4.7 we saw that the discontinuity of an amplitude across a cut associated with an n-particle normal threshold has the form

$$\text{Disc}_n \; \equiv \!\!\left(\!A\!\right)\!\!\equiv \quad = \quad \equiv\!\!\left(\!A_1\!\right)\!\!\diagdown\diagdown\!\!\left(\!A_2\!\right)\!\!\equiv \;\; , \qquad (4.10.1a)$$

or more concisely, $\qquad \text{disc}_n A = \int A_1 A_2 \, d\Omega_n. \qquad (4.10.1b)$

Here A_1 and A_2 represent certain boundary values of the amplitudes for the process involved and $d\Omega_n$ is the n-particle intermediate-state phase-space integral. According to the rules of (4.2.30) the latter is

$$\int \prod_{l=1}^{n-1} \left(\frac{i d^4 k_l}{(2\pi)^4}\right) \prod_{i=1}^{n} \left(-2\pi i \, \delta^{(+)}(q_i^2 - \mu_i^2)\right), \qquad (4.10.2)$$

where the $(n-1)$ loop momenta are chosen as in Fig. 4.10.1. The normal-threshold singularities in A_1 and A_2 will generate through the integral (4.10.1b) new singularities of $\mathrm{disc}_n A$. Since the latter is just the difference of A on two of its Riemann sheets these singularities must be singularities of A itself. Applying the same argument to discontinuities of A_1 and A_2 will produce new singularities of A_1 and A_2 also. These new singularities will generate through (4.10.1b) further singularities of A, and so on. In this way one can build up a series of successively more complicated stages in the singularity structure of amplitudes, the first stage being just the poles and normal thresholds. The minimum consistent singularity structure is the smallest set closed under this iteration.

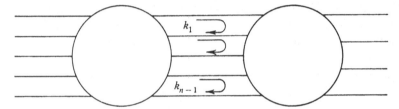

Fig. 4.10.1. The loop momenta in the n-particle unitarity integral.

We shall prove the result that the positions of the singularities generated in this way are given, just as in perturbation theory, by the solutions of the Landau equations corresponding to the set of Feynman-like diagrams that have appropriate external lines. In this context these diagrams are called *Landau–Cutkosky diagrams*, a name which serves to emphasise their S-matrix status, though they are identical in appearance to the set of Feynman diagrams of perturbation theory.

The proof proceeds by induction. The starting-point of the induction is provided by the normal thresholds, which certainly correspond to solutions of the Landau equations. The argument is completed by showing that singularities of A_1 and A_2 corresponding to solutions of the Landau equations will give a singularity of (4.10.1b) also having this property. We suppose therefore that A_1 has a singularity S_1 corresponding to a diagram with lines with internal momenta $q_i^{(1)}$, parameters $\alpha_i^{(1)}$, masses $\mu_i^{(1)}$, and with loop momenta $k_j^{(1)}$. The equation of S_1 is given by eliminating the $\alpha_i^{(1)}$ and $k_j^{(1)}$ from

$$\psi_1 \equiv \sum_i \alpha_i^{(1)}(q_i^{(1)2} - \mu_i^{(1)2}) = 0 \qquad (4.10.3)$$

by means of the equations

$$\frac{\partial \psi_1}{\partial \alpha_i^{(1)}} = 0, \quad \frac{\partial \psi_1}{\partial k_j^{(1)}} = 0. \qquad (4.10.4)$$

The resulting Landau curve has among its variables quantities depending on the loop momenta k_l of the phase space integration (4.10.2). We suppose that A_2 has a singularity S_2 defined in a similar way in terms of variables $q_i^{(2)}$, $\alpha_i^{(2)}$, $\mu_i^{(2)}$ and $k_j^{(2)}$.

The singularity which results from combining S_1 and S_2 in the integral (4.10.1) can be found by the methods of §2.1 with the aid of a slight modification. In that section we considered only integrands which were analytic apart from certain singularities. It is necessary, therefore, to replace the non-analytic δ-functions in (4.10.2) by factors $(q_i^2 - \mu_i^2)^{-1}$ together with the prescription that the hypercontour encircles each pole $q_0 = +\sqrt{(\mathbf{q}^2 + \mu^2)}$. By the residue theorem this is an equivalent integral. Since A_1 and A_2 are only defined *at* the poles the hypercontour must be taken infinitesimally close to each pole.†

The singularities will be given by

$$\left. \begin{aligned} \lambda_1 S_1 = 0, \quad \lambda_2 S_2 = 0, \\ \alpha_i(q_i^2 - \mu_i^2) = 0 \quad (i = 1, 2, ..., n), \\ \frac{\partial}{\partial k_l}[\lambda_1 S_1 + \lambda_2 S_2 + \Sigma \alpha_i(q_i^2 - \mu_i^2)] = 0 \quad (l = 1, 2, ..., n-1). \end{aligned} \right\} \qquad (4.10.5)$$

The k_l dependence of S_1 will arise in two distinct ways. Some of the $q_i^{(1)}$ in (4.10.3) will contain k_l explicitly; these will be just the lines in the diagram for S_1 which lie round the k_l-loop when the diagrams for S_1 and S_2 are joined together. In addition, the $\alpha_i^{(1)}$ and $k_j^{(1)}$ are dependent on the external variables of A_1 and so, in particular, depend on k_l. However, equations (4.10.4) imply that

$$\frac{\partial S_1}{\partial k_l} = \frac{\partial \psi_1}{\partial k_l} + \Sigma \frac{\partial \psi_1}{\partial \alpha_i^{(1)}} \frac{\partial \alpha_i^{(1)}}{\partial k_l} + \Sigma \frac{\partial \psi_1}{\partial k_j^{(1)}} \frac{\partial k_j^{(1)}}{\partial k_l}$$

$$= \frac{\partial \psi_1}{\partial k_l},$$

so that the third set of equations (4.10.5) can be written

$$\sum_{k_l \text{ loop}} [\lambda_1 \alpha_i^{(1)} q_i^{(1)} + \lambda_2 \alpha_i^{(2)} q_i^{(2)} + \alpha_i q_i] = 0. \qquad (4.10.6)$$

---

† A more sophisticated way to derive the result is to consider (4.10.2) as an integral defined in the manifold given by the equations $q_i^2 - \mu_i^2 = 0$. The permissible distortions of the hypercontour must always be within this manifold; compare the discussion of boundaries in §2.1. This leads to (4.10.5) with the α_i being Lagrange multipliers.

Equations (4.10.4–6) together are just the Landau equations associated with

$$\phi = \Sigma \tilde{\alpha}_i^{(1)}(q_i^{(1)^2} - \mu_i^{(1)^2}) + \Sigma \tilde{\alpha}_i^{(2)}(q_i^{(2)^2} - \mu_i^{(2)^2}) + \Sigma \alpha_i(q_i^2 - \mu_i^2), \quad (4.10.7)$$

where $\tilde{\alpha}_i^{(1)} = \lambda_1 \alpha_i^{(1)}$ and $\tilde{\alpha}_i^{(2)} = \lambda_2 \alpha_i^{(2)}$. This substantiates the statement that the singularity in a unitary integral generated by two singularities of Landau type is itself of Landau type.

This mechanism for the generation of singularities finds a very natural expression in graphical form since (4.10.7) just corresponds to the diagram formed by joining together the diagrams associated with S_1 and S_2. For example, if we have the singularities

then the integral formed from the two amplitudes has the singularity

and so on. Such a consideration provides the justification for the argument in §4.4 inferring the existence of the triangle singularity (4.4.3).

The conditions given for a singularity are necessary rather than sufficient. Sufficiency requires an analysis of whether the hypercontour is actually trapped and a consideration of possible cancellations. We shall go some way in a discussion of these points in the rest of this section. However, the necessary conditions alone permit us to make powerful negative statements that certain diagrams cannot have certain singularities. Thus cannot have the pole (4.5.3) since the insertion of diagrams into the bubbles, even allowing for contraction of the δ-function lines, cannot give rise to the pole. This type of result was used extensively in §§4.5 and 4.6.

Riemann sheet structure

We have seen that the Landau equations provide the possible locations of the singularities both of perturbation theory and of a unitary S-matrix theory. We also need to know the Riemann sheet structure of the singularities. In principle these are obtained by continuation from physical regions, singularities being found along those paths of continuation which lead to the hypercontour actually being trapped. In practice, however, this prescription is too difficult to use directly and we have to have recourse to indirect arguments. It is of interest to inquire to what extent we may expect the Riemann sheet properties of perturbation theory and a unitary S-matrix theory to coincide. In § 4.11 the explicit example of the triangle singularity will be considered in S-matrix theory and its physical-region properties are there found to be the same as in perturbation theory. Here we shall see to what extent the problem can be discussed in general.

We suppose that there is some limited region in the space of external invariants of an amplitude for which its analytic properties are the same in perturbation theory and S-matrix theory. That is to say the two theories have identical singularities in this region and the discontinuities around these singularities have the same structure in terms of other amplitudes. Then there will be an identical spectral representation for the amplitude in this region in the two theories. An example of this would be provided by the case where the two theories were known to satisfy the same dispersion relation for a range of fixed momentum transfer.

If now we continue from the region, new singularities may arise and we are concerned with the extent to which this will happen in an identical manner in the two theories. These singularities are generated in each term of the spectral representation by singularities of the discontinuities, which are themselves generated in the manner described earlier in this section. If we assume that there are no subtle cancellation mechanisms the presence or absence of singularities in each term is due to the trapping or not of the hypercontour. The distortions of the hypercontour are determined by the *location* of singularities in the integrand, but not by their nature (square root, logarithmic, etc.). To the extent therefore that the two theories have identical crossed channel singularities causing the distortions of the hypercontour they should lead to identical pinching configurations and identically located new singularities. If one considers the singularity structure of an S-matrix theory as being generated by successive iterations, starting with the normal threshold picture, then one could hope to use perturbation theory as a sort of mathematical probe to determine the occurrence of pinching configurations.

The utility of this notion is restricted by the fact that even at the first stage, namely the pole and normal threshold structure, finite-order perturbation theory and unitary S-matrix theory do not have identical singularity structures on unphysical sheets.

As an example of this consider the stable particle pole in the direct channel of the amplitude ⊐◯⊏. Suppose the unitarity relation

$$\text{⊐⊕⊏} - \text{⊐⊖⊏} = \text{⊐⊕⊏⊐⊖⊏} \qquad (4.10.8)$$

is continued to the left of the normal threshold $s = 4\mu^2$ in the total-energy variable s, so as to reach the pole $s = \mu^2$ of ⊐⊕⊏. This is the point P in Fig. 1.3.4, so that continuation must be taken to the upper side of the branch-point at $s = 4\mu^2$; this continuation takes ⊐⊖⊏ on to a different Riemann sheet. On this sheet there cannot be a pole at $s = \mu^2$, because then the continuation of the term on the right-hand side of (4.10.8) would have a double pole at $s = \mu^2$. But if the right-hand side of (4.10.8) has a double-pole, then so must at least one of the terms on the left. But this in turn would imply that the right-hand side had a pole of still higher order, and so on. Hence unitarity forbids the occurrence of the poles in the same place on different Riemann sheets, though in finite-order perturbation theory the poles are indelibly present on all sheets. To show their absence on some sheets requires the summation of an infinite set of Feynman diagrams. Similar considerations apply to certain normal thresholds (Landshoff, 1963; Olive, 1963a).

Another possible infinite-summation effect that should be mentioned is the occurrence of dense accumulations of singularities on a line, that is of natural boundaries of the analytic functions across which no continuation is possible (Freund & Karplus, 1961).

It is clear also that the continuation argument leaves unsettled questions relating to which singularities are assumed in the initial region. A famous problem here is the presence or absence of anomalous thresholds in two-particle → two-particle scattering amplitudes. However, some particular arguments can sometimes be given by considering consistency requirements. [Polkinghorne (1962d); see also § 4.11.]

Discontinuities

Since the Cutkosky formula (§ 2.9) can be written in a form which makes no reference to its perturbation-theory origin we are encouraged to suppose that it has a more general S-matrix basis. This we proceed to show in outline, using the picture of the generation of singularities given above.

We start with an amplitude in a given region which can be expressed in terms of a spectral representation containing terms like

$$\int_C ds' \frac{\mathrm{disc}_{S_1}[A(s')]}{(s'-s)},\qquad(4.10.9)$$

where S_1 is one of the singularities of A and $\mathrm{disc}_{S_1}[A]$ the corresponding discontinuity. As one continues from this region a new singularity, S_2, may arise in A from a pinch in (4.10.9) between the pole $(s'-s)^{-1}$ and the singularity S_2 present in $\mathrm{disc}_S[A(s')]$. If one encircles this

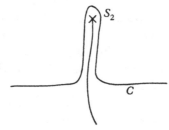

Fig. 4.10.2. The distortion of C by S_2.

singularity the contour C will be distorted by the singularity S_2 in $\mathrm{disc}_{S_1}[A]$, as shown in Fig. 4.10.2. This has the effect of producing an extra term in the spectral representation

$$\int_{C'} ds' \frac{\mathrm{disc}_{S_2}[\mathrm{disc}_{S_1}[A(s')]]}{s'-s},\qquad(4.10.10)$$

where the numerator in (4.10.10) is the discontinuity in $\mathrm{disc}_{S_1}[A(s')]$ on encircling S_2, and C' is a contour following the cut trailed by S_2. Equation (4.10.10) is clearly the extra term in the spectral representation due to the new singularity S_2 and this enables us to make the identification

$$\mathrm{disc}_{S_2}[A(s)] = \mathrm{disc}_{S_2}[\mathrm{disc}_{S_1}[A(s)]],\qquad(4.10.11)$$

a formula which was also given in perturbation theory by Cutkosky (1960). The merit of (4.10.11) is that it reduces the problem of finding discontinuities of amplitudes to the problem of finding discontinuities of unitarity-like integrals. This proves a simpler question.

It is simple to study discontinuities of one-dimensional integrals. Suppose in such an integral a singularity S is generated by a pinch between a singularity S_1 of a factor A_1 in the integrand and a singularity S_2 of a factor A_2. If we encircle S then in the integration complex plane S_1 and S_2 describe closed paths which impinge on the contour of integration. In the course of this movement S_1 and S_2 trail their cuts behind them. The initial and final configurations are illustrated in Fig. 4.10.3. The difference of these two integrals is an integral joining S_1 to S_2 in whose integrand A_1 is replaced by $\mathrm{disc}_{S_1}[A_1]$ and A_2 is replaced by $\mathrm{disc}_{S_2}[A_2]$. For multiple integrals this generalises to

$$\mathrm{disc}_S\left[\int A_1 A_2\, d\Omega_n\right] = \int \mathrm{disc}_{S_1}[A_1]\,\mathrm{disc}_{S_2}[A_2]\, d\Omega_n, \quad (4.10.12)$$

the boundary of the region of integration being given by parts of the singularity curves S_1 and S_2.

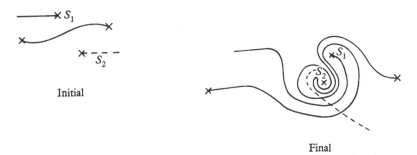

Fig. 4.10.3. The distortion of the contour when S is encircled.

It is clear from (4.10.12) that, if the discontinuities round S_1 and S_2 are given by the Cutkosky formula, so will be the discontinuity round S, since then the integral on the right-hand side will contain a mass shell δ-function for each line in the Landau–Cutkosky diagram corresponding to S. Since the iterative procedure starts with normal thresholds whose discontinuities are of Cutkosky type, it will only build up singularities whose discontinuities are also of this type.

This discussion is clearly an outline one, just as is the perturbation-theory argument. Neither determines the subtle question of the branches to be chosen for the amplitudes appearing in the Cutkosky formula. This requires more detailed study, of which a particular example is given in §4.11.

Further singularities

All the arguments we have given operate by iteration. If there are other singularities in addition to those we used as our starting-point, these will also participate in the iteration and so reproduce themselves in a similar way.

One such class of singularities which certainly exists is the set corresponding to unstable particles. As we saw in § 4.9, the only mathematical difference between stable and unstable particles is that the latter correspond to *complex* poles. So by carrying through the iteration starting with these poles, we see that we get an exactly similar set of singularities, corresponding to Feynman-like diagrams in which some or all of the internal lines are associated with complex mass. The Riemann-sheet properties of these singularities will differ from ones just involving stable particles because the complex poles do not occur on the physical sheet. The variable-force model discussed

(a)   (b)

Fig. 4.10.4. (a) Example of a singularity involving an internal unstable particle. (b) The singularity that (a) passes round in the transition to instability of that particle.

in § 4.9 predicts that in the transition from stability to instability of a particle of mass M the unstable particle singularity will pass round the branch-point obtained by replacing the M lines by pairs of m lines. For example the singularity represented by the diagram in Fig. 4.10.4a passes round that of Fig. 4.10.4b.

It is hoped that these, together with of course the second-type singularities of § 2.10, exhaust the set of singularities.

4.11 The triangle singularity in the physical region

In the previous section in the attempt to discuss Riemann sheet properties a new assumption was introduced, to the effect that in certain limited regions of its variables the amplitude obeys spectral representations like those known from perturbation theory. It is clearly highly desirable to be able, if possible, to do away with this assumption. We shall now see from an example that the physical unitarity equations, together with the relatively weak analyticity assumptions formulated in § 4.4, can unambiguously control the Riemann sheet structure, giving rise to the Cutkosky formulae of (2.9.10) (Landshoff & Olive, 1966).

We shall look first at singularities in the physical region, since it is here that the unitarity equations initially operate, and are easiest to analyse, and shall consider the simplest example, the singularity in the three-particle $\rightarrow$ three-particle amplitude whose Landau diagram takes the form of the triangle Fig. 4.11.1. The masses will be treated as

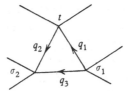

Fig. 4.11.1. The Landau diagram under study labelled with the variables used in the text.

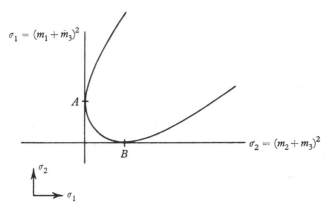

Fig. 4.11.2. The part of singularity curve L for the triangle diagram lying in the physical region with t fixed.

if unequal in order to keep track in subsequent manipulations (and show that we can cope with general kinematics) even though we work with equal-mass unitarity equations. If we fix t at a physical value ($t < 0$) the Landau curve L in the plane of the variables σ_1 and σ_2 is a hyperbola whose branch lying in the physical region is shown in Fig. 4.11.2. L touches the σ_1 and σ_2 normal thresholds at A and B. It is the arc AB which is singular in the physical limit according to the criteria of perturbation theory (§ 2.3).

We mentioned in § 4.4 that only three of the terms in the three-particle unitarity equation were expected to generate this singularity (see (4.4.3)). That no other terms in the equation have this property.

follows from the work of § 4.10, and so we can write the equation in the form

$$\equiv\!\!\bigodot\!\!{+}\!\!\equiv\; -\; \equiv\!\!\bigodot\!\!{-}\!\!\equiv\; =\; \equiv\!\!\bigcirc\!\!{+}\!\!\bigodot\!\!{-}\!\!\equiv\; +\; \equiv\!\!{+}\!\!\bigodot\!\!\bigcirc\!\!{-}\!\!\equiv\; +\; \equiv\!\!{+}\!\!\bigodot\!\!\equiv\!\!{-}\!\!\equiv$$

$$+ \ \text{(Terms regular on } L\text{).} \qquad (4.11.1)$$

The singularity L is also regenerating; it occurs in each of the first two terms on the right-hand side of (4.11.1) as a direct result of its occurring in the amplitudes $\equiv\!\bigcirc\!\equiv$ appearing within those terms.

We shall examine in detail how the terms in (4.11.1) generate L by the mechanisms (4.4.3) and use the results to calculate from (4.11.1) the discontinuity of $\equiv\!\!(+)\!\!\equiv$ across L by methods like those of §§ 4.5 and 4.6. The result is

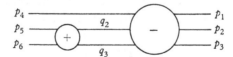

$$\left.\begin{array}{ll} \text{Disc}\ \equiv\!\!(+)\!\!\equiv\ = & \quad\text{on } AB, \\[2mm] 0 & \quad\text{on } \infty A \text{ and } B\infty. \end{array}\right\} \qquad (4.11.2)$$

The integral is to be interpreted by the rules (4.2.30). Thus we confirm the perturbation-theory result that $\equiv\!\!(+)\!\!\equiv$ is only singular on AB, where its discontinuity is given by a Cutkosky formula. Note that by studying the physical unitarity formula we have made a return to first principles and that the result is hence rigorous, making no use of crossing or hermitian analyticity.

Analysis of the integrals†

Let us label the lines of the first term on the right-hand side of (4.11.1) as shown in Fig. 4.11.3. This choice conforms with that in Fig. 4.11.1. In order to treat the integral by standard methods we integrate out the δ-functions corresponding to each internal line and to do this we transform to the invariant variables (4.3.16).

Fig. 4.11.3. The momenta of the integral under study.

† Alternative methods of analysis, which do not make use of Cayley and Gram determinants, are given by Landshoff, Olive & Polkinghorne (1966a) and Boyling (1966).

In the integral the p's, being external vectors, are fixed, while the two q's vary over all possible real values subject to the energy-momentum conservation and mass shell constraints. So the integration region in the new variables (which is all we need for the present purpose) can be found from the conditions (4.3.13) for the construction of the sequence of vectors
$$p_4, \quad p_1, \quad p_2 + p_3, \quad q_3.$$

The first four conditions are physical-region constraints for the $(-)$-amplitude while the last, using the trick (4.3.18), can be written
$$G(q_3, q_3 - p_2 - p_3, q_3 - p_1 - p_2 - p_3, q_3 + p_4 - p_1 - p_2 - p_3) < 0$$

or, by (4.3.17),

$$M(u_1, v) = \begin{vmatrix} 0 & 1 & 1 & 1 & 1 & 1 \\ 1 & 0 & \sigma_1 & s & \sigma_2 & \mu_3^2 \\ 1 & \sigma_1 & 0 & M_1^2 & t & u_1 \\ 1 & s & M_1^2 & 0 & M_4^2 & v \\ 1 & \sigma_2 & t & M_4^2 & 0 & \mu_2^2 \\ 1 & \mu_3^2 & u_1 & v & \mu_2^2 & 0 \end{vmatrix} > 0, \qquad (4.11.3)$$

where $p_i^2 = M_i^2$, $q_i^2 = \mu_i^2$, $s = (p_1 + p_2 + p_3)^2$, $u_1 = (q_3 - p_2 - p_3)^2$,

$v = (p_4 + q_2)^2$ and $\sigma_1 = (p_2 + p_3)^2$, $\sigma_2 = (p_5 + p_6)^2$, $t = (p_1 - p_4)^2$

as indicated in Fig. 4.11.1. The variables u_1 and v replace q_1 and q_2 as integration variables, being integrated over the region (4.11.3).

The singularity of the integrand indicated by (4.4.3a) is the pole $S \equiv u_1 - \mu_1^2 = 0$. In this case the conditions for singularity reduce to

$$M = \frac{\partial M}{\partial v} = 0, \qquad (4.11.4)$$

with $u_1^2 = \mu_1^2$. This may be regarded as an end-point singularity of the u_1 integration when the v integration has been carried out. From (4.11.3),
$$\frac{\partial M}{\partial v} = 2M_5^3,$$

where M_j^i is the (i,j) algebraic minor of the Cayley determinant M (with rows and columns numbered $0, 1, \ldots, 5$ as indicated in (4.3.17)). Jacobi's theorem (4.3.14) takes the form

$$M_j^i M_l^k - M_l^i M_j^k = M_{jl}^{ik} M. \qquad (4.11.5)$$

Putting $i = j = 3$ and $k = l = 5$, we see that (4.11.4) implies that either $M_3^3 = 0$ or $M_5^5 = 0$. The first possibility is the one of interest here since it involves u_1 (the other yields a second-type singularity (see §2.10 and also (4.3.20)). It is

$$R(\mu_1^2, \mu_2^2) = 0, \tag{4.11.6}$$

where

$$R(u_1, \mu_2^2) = M_3^3 = \begin{vmatrix} 0 & 1 & 1 & 1 & 1 \\ 1 & 0 & \sigma_1 & \sigma_2 & \mu_3^2 \\ 1 & \sigma_1 & 0 & t & u_1 \\ 1 & \sigma_2 & t & 0 & \mu_2^2 \\ 1 & \mu_3^2 & u_1 & \mu_2^2 & 0 \end{vmatrix}. \tag{4.11.7}$$

By (4.3.17), (4.11.6) can be written in the form $G(-q_3, q_1, q_2) = 0$, which is indeed the Landau equation for L (see (2.3.6)).

We calculate the discontinuity of the integral in Fig. 4.11.3, arising from this mechanism of generating the triangle singularity, by fixing t and taking σ_1 and σ_2 around L on a path lying close to L in a plane normal to L in the four-dimensional complex (σ_1, σ_2)-space. Thus on the path the displacement of (σ_1, σ_2) from L is given by

$$d\sigma_1 = d\eta \left[\frac{\partial}{\partial \sigma_1} R(\mu_1^2, \mu_2^2) \right], \quad d\sigma_2 = d\eta \left[\frac{\partial}{\partial \sigma_2} R(\mu_1^2, \mu_2^2) \right], \tag{4.11.8}$$

with the derivatives evaluated on L. It may readily be ascertained that a real positive $d\eta$ corresponds to a displacement along the inward normal to L. For a variation in u_1, and in σ_1 and σ_2 of the type (4.11.8),

$$dR = Q d\eta + \left[\frac{\partial}{\partial u_1} R(u_1, \mu_2^2) \right]_{u_1 = \mu_1^2} du_1, \tag{4.11.9}$$

where

$$Q = \left[\frac{\partial}{\partial \sigma_1} R(\mu_1^2, \mu_2^2) \right]^2 + \left[\frac{\partial}{\partial \sigma_2} R(\mu_1^2, \mu_2^2) \right]^2.$$

The displacement of the integration end-point from $u_1 = \mu_1^2$ is given by $dR = 0$, so it is

$$du_1 = \frac{-Q d\eta}{\partial R / \partial u_1}. \tag{4.11.10}$$

In terms of cofactors of the determinant (4.11.7)

$$\partial R / \partial u_1 = 2 R_4^2,$$

and Jacobi's identity (4.11.5) applied to R on L gives

$$(R_4^2)^2 = R_2^2 R_4^4.$$

Hence $du_1 / d\eta$ changes sign when either R_2^2 or R_4^4 vanishes. The latter possibility does not occur on the branch of L we are discussing (since $t < 0$); the former occurs at the normal threshold $\sigma_2 = (\mu_2 + \mu_3)^2$, that is at B in Fig. (4.11.2). So $du_1/d\eta$ takes different signs on the arcs ∞AB and $B\infty$ in the figure, and it is easy to see that it is positive on the former and negative on the latter, by calculating it at A using the fact that L always touches the σ_1 normal threshold which itself increases with m_1. Using Jacobi's theorem, and the fact that Cayley determinants of real external vectors are positive in the physical region, it follows that outside L in the real σ_1, σ_2 plane, $M < 0$ when $u_1 = \mu_1^2$, so that the point $u_1 = \mu_1^2$ lies inside the integration region only for points inside L. Hence the orientation of the contour in the u_1-plane for variations in η is as shown in Fig. 4.11.4.

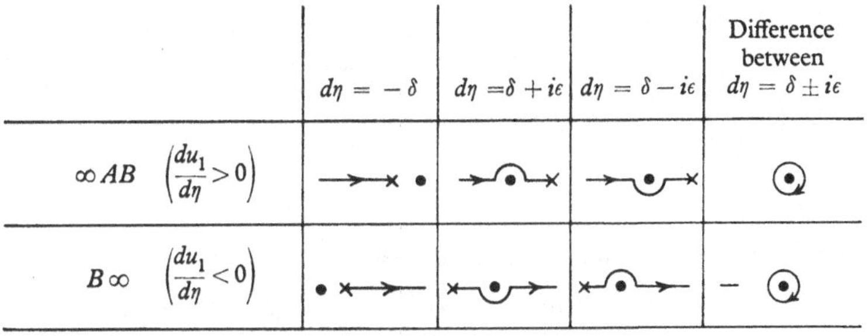

	$d\eta = -\delta$	$d\eta = \delta + i\epsilon$	$d\eta = \delta - i\epsilon$	Difference between $d\eta = \delta \pm i\epsilon$
∞AB $\left(\dfrac{du_1}{d\eta} > 0\right)$	→×•	→⌢×	→⌣×	⊙
$B\infty$ $\left(\dfrac{du_1}{d\eta} < 0\right)$	•×→	×⌣→	×⌢→	— ⊙

Fig. 4.11.4. Table representing the orientation of the u_1-integration contour in the integral in Fig. 4.11.3 with respect to the pole $u_1 = m_1^2$. The column $d\eta = -\delta$ corresponds to points (σ_1, σ_2) outside L, the next two columns correspond to continuations to points inside L by paths with Im $\eta > 0$, Im $\eta < 0$ respectively.

It follows from the last column that the discontinuity in the variable η of the integral, across L, is obtained by replacing the factor with the pole by the residue of that pole times the factor $\pm 2\pi i\, \delta(u_1 - \mu_1^2)$, with the sign corresponding to arcs $B\infty$ and ∞AB respectively. Transforming back from the integration over invariants to the loop integration we see that the answer can be written according to the rules (4.2.30) as

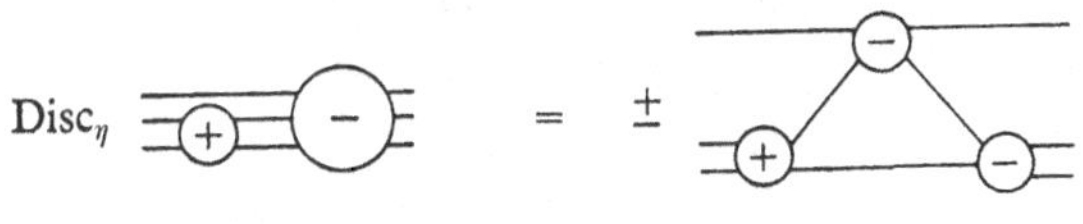

$$\text{Disc}_\eta \quad \boxed{+}\;\boxed{-} \quad = \quad \pm \quad \triangle \qquad , \quad (4.11.11a)$$

with $+$ on ∞AB and $-$ on $B\infty$. The subscript η is to emphasise that the discontinuity is taken in the variable η. This result shows that the integral is singular all round L.

Similarly, we can analyse the process (4.4.3b) and find that

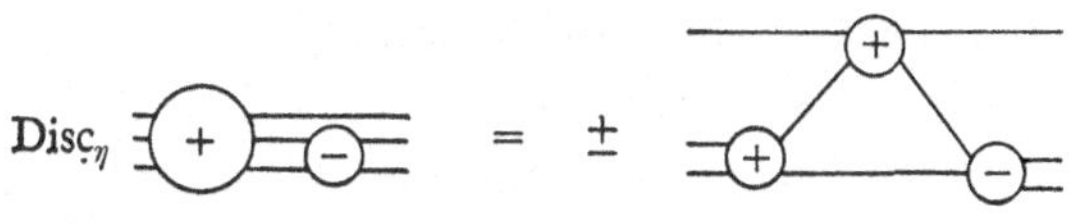

$$\text{Disc}_\eta \quad \boxed{+}\;\boxed{-} \quad = \quad \pm \quad \triangle \qquad , \quad (4.11.11b)$$

with + on $AB\infty$ and − on ∞A. In obtaining (4.11.11 a, b) we have ignored the regeneration effect mentioned earlier, whereby the integrals are singular on L because ⊒◯⊏ are. We take account of this in the final stage of the argument.

The third term on the right-hand side of (4.11.1),

is more difficult to analyse since the process (4.4.3c) involves two singularities. It can be shown that this integral is singular only on the arcs ∞A and $B\infty$, and furthermore that its discontinuity is given by (see Landshoff & Olive, 1966)

$$\text{Disc}_\eta\ \sqsupseteq\!\!\bigcirc\!\!+\ \sqsupset\!\!\sqsubset\ -\ \bigcirc\!\!\sqsupset\ =\ \begin{pmatrix}+1\\0\\-1\end{pmatrix}\ \ \ \ \ \text{on}\ \begin{pmatrix}\infty A\\AB\\B\infty\end{pmatrix}.$$

(4.11.11c)

Comparison of the unitarity terms on either side of L

Contrary to what one might expect at first sight, the two versions of the unitarity equation (4.11.1) valid inside and outside L are independent equations. As we shall now see, this is because the terms on the right-hand side of the two equations are not always continuations of themselves. The comparison of the two equations will eventually enable us to calculate the discontinuity of ⊒◯+⊏ across L, but first we must compare each individual term of (4.11.1).

So far we do not know whether or not ⊒◯+⊏ is singular on the whole of L, so that until we prove otherwise we shall suppose that all of L is singular. According to our analyticity assumption there is to be some way of continuing around L to analytically relate ⊒◯+⊏ as defined on either side. Since the variable η of (4.11.8), being measured along the inward normal, agrees in sense with σ_1 and σ_2 at A and B respectively, the argument of §4.4 indicates that L must have an $\eta + i\epsilon$ distortion (see Fig. 4.11.5) since, by assumption, the σ_1 and σ_2 normal thresholds have $\sigma_1 + i\epsilon$ and $\sigma_2 + i\epsilon$ distortions respectively. Similarly (or just by complex conjugation) we would obtain an $(\eta - i\epsilon)$ rule for the continuation linking the physical ⊒◯−⊏ amplitude.

When the first term on the right-hand side of (4.11.1) is evaluated inside L (the region $\eta > 0$), the label (−) on the right-hand bubble indicates that the integration contour in the u_1-plane is depressed below the pole $u_1 = m_1^2$. From Fig. 4.11.4 we see that such a contour can only be obtained by continuing the integral defined outside L ($\eta < 0$) with an ($\eta + i\epsilon$) path when the arc $B\infty$ is traversed and with an ($\eta - i\epsilon$) path when the arc ∞AB is traversed. We say that distortions $\eta - i\epsilon$ and $\eta + i\epsilon$ are 'natural' on ∞AB and $B\infty$ respectively. 'Natural distortions' for the continuation of the other two terms can also be determined (see Landshoff & Olive, 1966) and are tabulated in Fig. 4.11.6.

So far we have mentioned only the generation of the triangle singularity L by the poles. We must also remember that L also occurs in each of the first two terms on the right-hand side of (4.11.1) as a result of its occurrence within the amplitudes ⊐○⊏ appearing in those terms. Accordingly, in the continuation of either of these terms across L the amplitudes within the integrals must be continued along the same path as the term as a whole.

Calculation of the discontinuity in ⊐(+)⊏ *across* L

We now have enough information to continue the unitarity equation valid outside L into the region inside L. We shall follow an ($\eta - i\epsilon$) path, so that ⊐(−)⊏ is continued into ⊐(−)⊏ while ⊐(+)⊏ is

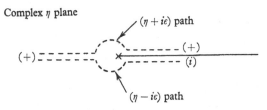

Fig. 4.11.5. Paths ($\eta \pm i\epsilon$) connecting real points outside L to those inside L.

continued into a new region (i) separated from the physical (+) boundary-value by the cut whose discontinuity we seek (see Fig. 4.11.5). This new equation will relate ⊐(i)⊏ and ⊐(−)⊏, in contradistinction to the physical unitarity equation operating inside L, which relates ⊐(+)⊏ and ⊐(−)⊏.

We first suppose that the arc $B\infty$ is traversed. Fig. 4.11.6 shows that the ($\eta - i\epsilon$) distortion is not the natural one for any of the terms on the right-hand side of (4.11.1). We put this right by using the dis-

continuity formulae (4.11.11), so finding for the continuation of
(4.11.1)

$$\equiv\!\bigcirc\!i\!\equiv\; - \;\equiv\!\bigcirc\!-\!\equiv \;=\; \left[\equiv\!\!\oplus\!\equiv\!\bigcirc\!-\!\equiv \;+\; \triangle \right] \;+$$

$$\left[\equiv\!\bigcirc\!i\!\equiv\!\ominus\!\equiv \;-\; \triangle \right] \;+\; \left[\equiv\!\oplus\!\equiv\!\bigcirc\!-\!\equiv \;+\; \triangle \right]$$

+(terms regular on L).   (4.11.12)

Using the two-particle unitarity equation

$$\equiv\!\oplus\!\equiv \;-\; \equiv\!\ominus\!\equiv \;=\; \equiv\!\oplus\!\ominus\!\equiv \;=\; \equiv\!\ominus\!\oplus\!\equiv, \quad (4.11.13)$$

we find that the second, fourth and sixth terms on the right-hand side
of (4.11.12) cancel. Subtracting the result from the version of (4.11.1)
valid inside L, and rearranging, we have

$$\left(\equiv\!\oplus\!\equiv \;-\; \equiv\!\bigcirc\!i\!\equiv\right)\!\left(\equiv\!\equiv \;-\; \equiv\!\ominus\!\equiv\right) \;=\; 0. \quad (4.11.14)$$

'Postmultiplying' this equation by the expression

$$\equiv \;+\; \equiv\!\oplus\!\equiv, \quad (4.11.15)$$

and again using (4.11.13) we obtain

$$\equiv\!\oplus\!\equiv \;-\; \equiv\!\bigcirc\!i\!\equiv \;=\; 0. \quad (4.11.16)$$

Hence the arc $B\infty$ is not a singularity of $\equiv\!\oplus\!\equiv$.

The $(\eta - i\epsilon)$ continuation across the arc AB is unnatural only for the
second term on the right-hand side of (4.11.1) (according to Fig.
4.11.6), and so we find, using (4.11.11b)

$$\equiv\!\bigcirc\!i\!\equiv \;-\; \equiv\!\bigcirc\!-\!\equiv \;=\; \equiv\!\oplus\!\bigcirc\!-\!\equiv \;+\; \left[\equiv\!\bigcirc\!i\!\ominus\!\equiv \;-\; \triangle \right] \;+\; \equiv\!\oplus\!\bigcirc\!-\!\equiv$$

+(Terms regular on L).

Subtracting this from (4.11.1) and rearranging, we have

$$\left(\equiv\!\oplus\!\equiv \;-\; \equiv\!\bigcirc\!i\!\equiv\right)\!\left(\equiv\!\equiv \;-\; \equiv\!\ominus\!\equiv\right) \;=\; \triangle .$$

Postmultiplying by expression (4.11.15) and using (4.11.13) as before, we obtain finally

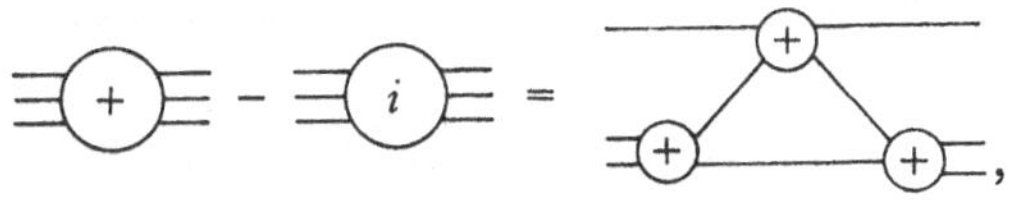

which is the formula predicted by Cutkosky.

Repetition of the procedure on ∞A yields (4.11.14) and hence (4.11.16), so that our final result can indeed be expressed in the form (4.11.2). Corresponding results for 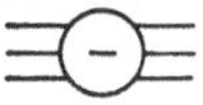 can be found by considering an $(\eta + i\epsilon)$ path of continuation instead.

Integral \ Arc of L	∞A	AB	$B\infty$
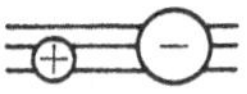	$\eta - i\epsilon$	$\eta - i\epsilon$	$\eta + i\epsilon$
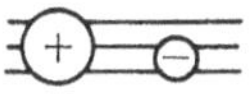	$\eta - i\epsilon$	$\eta + i\epsilon$	$\eta + i\epsilon$
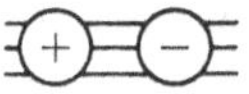	$\eta - i\epsilon$	Either	$\eta + i\epsilon$

Fig. 4.11.6. Table of paths of continuation from points outside L to points inside L giving the 'natural arrangement' of contours.

The analysis given applies only when (4.11.1) is valid, and in particular when σ_1 and σ_2 are each less than $(3\mu)^2$, so that at higher energies more complicated equations must be studied. We could, however, extend the result by analytic continuation of the above formulae to higher energies. That the two methods should agree would be yet another consistency requirement.

That the amplitude was singular only on AB was a special case of a general criterion for the singularity of Landau curves in the physical region, the 'positive-α criterion' known from perturbation theory. It is possible that the method above can be extended to prove this result in general, together with the associated discontinuity formulae.† If so this would be important for the following three reasons.

† *Note added in proof.* Further examples have been established by Bloxham (1966) and Landshoff, Olive & Polkinghorne (1966b).

(a) It would show that the singularity structure in the physical regions of the amplitudes is uniquely determined by unitarity and the requirement that the amplitudes be boundary values of analytic functions.

(b) It means that the amplitudes are only singular when there can be a physical intermediate scattering process. Since the positive-α criterion corresponds to the requirement that the intermediate particles move forward in time (see §2.5), it expresses an S-matrix notion of *causality*, that signals cannot be received before transmission.

(c) The physical region singularities would have the 'hierarchical property'—any given singularity curve stops being singular only when one (or more) of the α's vanishes. Then the curve touches the lower-order curve corresponding to the scattering process in which one (or more) less particles participate (see §2.6).

Anomalous thresholds

By the arguments of §4.10, the two-particle unitarity integral has the singularity

$$ \text{⊐⊕⊐⊖⊏} \;\longrightarrow\; \text{⊐◯⊏}. $$

Thus by (4.11.13) either ⊐⊕⊏ or ⊐⊖⊏ possesses the singularity, but the equation does not tell us which. We now outline how the $(+\alpha)$ criterion known in perturbation theory for the appearance of this 'anomalous threshold' on the physical sheet (see §2.3) follows from our previous results.

The factorisation in t and σ_1

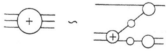

means that singularities of ⊐⊕⊏ imply ones in ⊒⊕⊒. But we know the singular behaviour of the triangle singularity in the physical region of ⊒⊕⊒ and can determine it elsewhere by tracing paths along the attached complex surfaces. In particular the methods of §2.3 imply that for $(2\mu)^2 > t > 0$ the singular arc AB of L in the (σ_1, σ_2)-plane is like that shown in Fig. 2.3.3a (though the variables are different the picture is the same). Looking at the mass shell values of t and σ_1 we find the desired result.

To see how the presence of anomalous thresholds affects the derivation of hermitian analyticity in §4.6 we must consider a theory with particles of different masses. Let particles of mass μ_1, m and M

correspond to straight, dotted and dashed lines respectively and suppose that the $\mu+\mu$ system is coupled to the lighter $m+m$ system (which is the only one of lower mass). To prove hermitian analyticity for ⊃⊕⊂ by means of the pole term (4.6.11), as before, we must compare the unitarity equations for ≡⟨ + ⟩≡ valid in the regions $s_{1235} < (2m)^2$ and $s_{1235} > (2\mu)^2$. These will have more intermediate states than (4.6.12) and (4.6.21) and we shall not write them down.

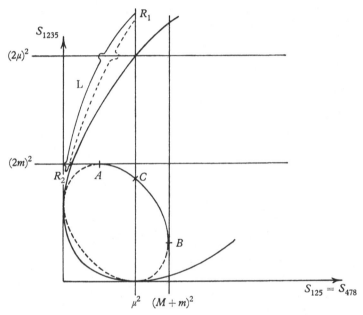

Fig. 4.11.7. This refers to the four-particle → four-particle amplitude labelled as in equation (4.6.11). The parabola bounds the physical region and the ellipse is the Landau curve for the triangle diagram in Fig. 4.11.8.

The argument goes as before until we consider the path Γ from R_1^+ to R_2^+ to R_1^i and how it can be distorted to lie in the mass shell section $s_{125} = s_{478} = \mu^2$. Instead of the situation in Fig. 4.6.4 we have that in Fig. 4.11.7.

The singular arc AB is found by the methods previously described and corresponds to the Landau diagrams in Fig. 4.11.8. If M is such that ⊃⊕⊂ has the anomalous threshold ⊃O⋯⊏ on the physical sheet then AB will intersect $s_{125} = s_{478} = \mu^2$ at C as shown.†

   † Because of the equal masses this could not happen to any corresponding arc inserted into Fig. 4.6.4.

As the path Γ is distorted to the mass-shell section it must not cut any singular curves and so it will be 'caught up' and stretched by the arc AB so that in the mass shell section it must encircle the singularity at C as well as the normal thresholds $(2m)^2$ and $(2\mu)^2$. In terms of the middle bubble of (4.6.11) this means that the anomalous threshold must be encircled as well as the normal thresholds in order to reach the hermitian analytic boundary value.

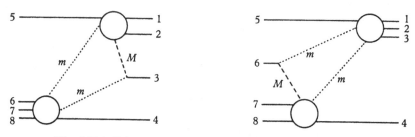

Fig. 4.11.8. Triangle diagrams whose Landau curves correspond to the ellipse in Fig. 4.11.7.

To prove extended unitarity for $\rightleftharpoons\!\!\!\!\bigoplus\!\!\!\!\rightleftharpoons$ in the range

$$(2\mu)^2 > s > (2m)^2$$

the unitarity equation for $\equiv\!\!\!\bigoplus\!\!\!\equiv$ in the corresponding range of s_{1235} would also have to be considered.

Boyling (1964b) has presented these arguments in greater detail, and shown by majorisation techniques that no essentially new difficulties arise from higher Landau curves.

REFERENCES

AITKEN, A. C. (1954). *Determinants and Matrices* (Oliver and Boyd).
AMATI, D., FUBINI, S. & STANGELLINI, A. (1962). *Nuovo Cimento*, **26**, 896.
ASRIBEKOV, V. (1962). *Soviet Physics—JETP*, **15**, 394.
AZIMOV, YA. I. (1963). *Phys. Lett.* **3**, 195.
BALL, J. S., FRAZER, W. R. & NAUENBERG, M. (1962). *Phys. Rev.* **128**, 478.
BETHE, H. A. & SALPETER, E. E. (1951). *Phys. Rev.* **84**, 1232.
BJORKEN, J. D. (1959). Stanford University Ph.D. thesis.
BJORKEN, J. D. & WU, T. T. (1963). *Phys. Rev.* **130**, 2566.
BLANKENBECLER, R., GOLDBERGER, M. L., MACDOWELL, S. W. & TREIMAN, S. B. (1961). *Phys. Rev.* **123**, 692.
BLOXHAM, M. J. W. (1966). *Nuovo Cimento* (to be published).
BOYLING, J. B. (1963). *Ann. Phys.* **25**, 249.
BOYLING, J. B. (1964a). *Ann. Phys.* **28**, 435.
BOYLING, J. B. (1964b). *Nuovo Cimento*, **33**, 1356.
BOYLING, J. B. (1966). *Nuovo Cimento* (to be published).
BRANSON, D. (1964). *Phys. Rev.* **135** B, 1255.
BRANSON, D. (1965). *Ann. Phys.* **35**, 351.
BYERS, N. & YANG, C. N. (1964). *Rev. Mod. Phys.* **36**, 595.
CASSANDRA, M., CINI, M., JONA-LASINIO, G. & SERTORIO, L. (1963). *Nuovo Cimento*, **28**, 1351.
CHALLIFOUR, J. & EDEN, R. J. (1963). *J. Math. Phys.* **4**, 359.
CHENG, H. & WU, T. T. (1965). *Phys. Rev.* **140** B, 465.
CHEW, G. F. (1962). *S-Matrix Theory of Strong Interactions* (Benjamin, New York).
CHEW, G. F., GOLDBERGER, M. L., LOW, F. E. & NAMBU, Y. (1957). *Phys. Rev.* **106**, 1337.
CHEW, G. F. & JACOB, M. (1964). *Strong Interaction Physics* (Benjamin, New York).
CHISHOLM, J. S. R. (1952). *Proc. Camb. Phil. Soc.* **48**, 300.
COLEMAN, S. & NORTON, R. E. (1965). *Nuovo Cimento*, **38**, 438.
CONTOGOURIS, A. P. (1965). *Nuovo Cimento*, **36**, 250.
CORDES, J. G. (1965). *Nuovo Cimento*, **39**, 157.
COURANT, R. & HILBERT, D. (1953). *Methods of Mathematical Physics*, vol. 1, chapter 3 (New York).
CUTKOSKY, R. E. (1960). *J. Math. Phys.* **1**, 429.
DIRAC, P. A. M. (1958). *Principles of Quantum Mechanics* (4th edition) (Oxford University Press).
DRUMMOND, I. T. (1963). *Nuovo Cimento*, **29**, 720.
EDEN, R. J. (1952). *Proc. Roy. Soc.* A, **210**, 388.
EDEN, R. J. (1960a). *Phys. Rev.* **119**, 1763.
EDEN, R. J. (1960b). *Proc. 1960 Conference on High Energy Physics, University of Rochester* (Interscience), p. 219.

EDEN, R. J. (1960c). *Phys. Rev.* **120**, 1514.

EDEN, R. J. (1960d). *Phys. Rev. Lett.* **5**, 213.

EDEN, R. J. (1961). *Phys. Rev.* **121**, 1567.

EDEN, R. J. & LANDSHOFF, P. V. (1964). *Phys. Rev.* **136** B, 1817.

EDEN, R. J. & LANDSHOFF, P. V. (1965). *Ann. Phys.* **31**, 370.

EDEN, R. J., LANDSHOFF, P. V., POLKINGHORNE, J. C. & TAYLOR, J. C. (1961a). *Phys. Rev.* **122**, 307.

EDEN, R. J., LANDSHOFF, P. V., POLKINGHORNE, J. C. & TAYLOR, J. C. (1961b). *J. Math. Phys.* **2**, 656.

EDEN, R. J. & TAYLOR, J. R. (1964). *Phys. Rev.* **133** B, 1575.

ERDELYI, A. *et al.* (1953). *Higher Transcendental Functions*, Vol. I (McGraw Hill, New York).

FAIRLIE, D. B., LANDSHOFF, P. V., NUTTALL, J. & POLKINGHORNE, J. C. (1962a). *J. Math. Phys.* **3**, 544.

FAIRLIE, D. B., LANDSHOFF, P. V., NUTTALL, J. & POLKINGHORNE, J. C. (1962b). *Phys. Lett.* **3**, 55.

FEDERBUSH, P. G. & GRISARU, M. T. (1963). *Ann. Phys.* **22**, 263, 299.

FOTIADI, D., FROISSART, M., LASCOUX, J. & PHAM, F. (1964). (Unpublished). See also FOTIADI *et al. Topology*, **4**, 159 (1965).

FOWLER, M. (1962). *J. Math. Phys.* **3**, 936.

FOWLER, M., LANDSHOFF, P. V. & LARDNER, R. W. (1960). *Nuovo Cimento*, **17**, 936.

FRAUTSCHI, S. (1963). *Regge Poles and S-Matrix Theory* (Benjamin, New York).

FRAZER, W. R. & FULCO, J. R. (1959). *Phys. Rev. Lett.* **2**, 365.

FRAZER, W. R. & FULCO, J. R. (1960a). *Phys. Rev.* **117**, 1603.

FRAZER, W. R. & FULCO, J. R. (1960b). *Phys. Rev.* **119**, 1420.

FREUND, P. G. & KARPLUS, R. (1961). *Nuovo Cimento*, **21**, 519.

FROISSART, M. (1961a). Proceeding of La Jolla Conference (unpublished).

FROISSART, M. (1961b). *Phys. Rev.* **123**, 1053.

FROISSART, M. (1964). *Rendiconti della Scuola Internazionale di Fisica 'E. Fermi'*, Corso XXIX (Academic Press, New York, ed. Wigner, E. P.).

GELL-MANN, M. & GOLDBERGER, M. L. (1962). *Phys. Rev. Lett.* **9**, 275.

GELL-MANN, M., GOLDBERGER, M. L., LOW, F. E., MARX, E. & ZACHARIASEN, F. (1964). *Phys. Rev.* **133** B, 145.

GELL-MANN, M., GOLDBERGER, M. L., LOW, F. E., SINGH, V. & ZACHARIASEN, F. (1964). *Phys. Rev.* **133** B, 161.

GELL-MANN, M., GOLDBERGER, M. L., LOW, F. E. & ZACHARIASEN, F. (1963). *Phys. Lett.* **4**, 265.

GELL-MANN, M., GOLDBERGER, M. L. & THIRRING, W. (1954). *Phys. Rev.* **95**, 1612.

GOLDBERGER, M. L. & WATSON, K. F. (1964a). *Collision Theory* (Wiley, New York).

GOLDBERGER, M. L. & WATSON, K. F. (1964b). *Phys. Rev.* **136** B, 1472.

GREENMAN, J. V. (1965). *J. Math. Phys.* **6**, 660.

GRIBOV, V. N. (1962). *Soviet Physics—JETP*, **14**, 1395.

GRIBOV, V. N. & POMERANCHUK, I. YA. (1962a). *Phys. Lett.* **2**, 239.

GRIBOV, V. N. & POMERANCHUK, I. YA. (1962b). *Phys. Rev. Lett.* **9**, 238.
GRIBOV, V. N., POMERANCHUK, I. YA. & TER MARTIROSYAN, K. A. (1965). *Phys. Rev.* **139**B, 184.
GUNSON, J. (1965). *J. Math. Phys.* **6**, 827, 845, 852.
GUNSON, J. & TAYLOR, J. G. (1960). *Phys. Rev.* **119**, 1121.
HADAMARD, J. (1898). *Acta Math.* **22**, 55.
HALLIDAY, I. G. (1963). *Nuovo Cimento*, **30**, 177.
HALLIDAY, I. G. (1964). *Annals of Physics*, **28**, 320.
HALLIDAY, I. G. & POLKINGHORNE, J. C. (1963). *Phys. Rev.* **132**, 2741.
HAMILTON, J. (1959). *Reports on Progress in Nuclear Physics*, **7**.
HAMILTON, J. (1964). *Strong Interactions and High Energy Physics* (Oliver and Boyd, ed. Moorhouse, R. G.).
HAMPRECHT, B. (1965). *Nuovo Cimento*, **40**, 542.
HEARN, A. C. (1961). *Nuovo Cimento*, **21**, 333.
HEISENBERG, W. (1943). *Z. Phys.* **120**, 513, 673.
HEISENBERG, W. (1944). *Z. Phys.* **123**, 93.
HWA, R. C. (1964). *Phys. Rev.* **134**B, 1086.
IAGOLNITZER, D. (1965). *J. Math. Phys.* **6**, 1576.
ISLAM, J. N. (1963). *J. Math. Phys.* **4**, 872.
ISLAM, J. N., LANDSHOFF, P. V. & TAYLOR, J. C. (1963). *Phys. Rev.* **130**, 2560.
IVANTER, I. G., POPOVA, A. M. & TER MARTIROSYAN, K. A. (1964). *Soviet Physics—JETP*, **19**, 568.
KÄLLÈN, G. & WIGHTMAN, A. S. (1958). *Dan. Vid. Selsk. Mat-fys. Skr.* **1**, no. 6.
KARPLUS, R., SOMMERFIELD, C. M. & WICHMANN, E. H. (1958). *Phys. Rev.* **111**, 1187.
KARPLUS, R., SOMMERFIELD, C. M. & WICHMANN, E. H. (1959). *Phys. Rev.* **114**, 376.
KASCHLUN, F. & ZOELLNER, W. (1965). *Nuovo Cimento*, **34**, 1618.
KENNEDY, J. & SPEARMAN, T. D. (1962). *Phys. Rev.* **126**, 1596.
KIBBLE, T. W. B. (1960). *Phys. Rev.* **117**, 1159.
KIBBLE, T. W. B. (1963). *Phys. Rev.* **131**, 2282.
LANDAU, L. D. (1959a). Proceedings of the Kiev Conference on High Energy Physics (unpublished).
LANDAU, L. D. (1959b). *Nuclear Phys.* **13**, 181.
LANDSHOFF, P. V. (1960). *Nuclear Phys.* **20**, 129.
LANDSHOFF, P. V. (1962). *Phys. Lett.* **3**, 116.
LANDSHOFF, P. V. (1963). *Nuovo Cimento*, **28**, 123.
LANDSHOFF, P. V. & OLIVE, D. I. (1966). *J. Math. Phys.* (to be published).
LANDSHOFF, P. V., OLIVE, D. I. & POLKINGHORNE, J. C. (1966a). *J. Math. Phys.* (to be published).
LANDSHOFF, P. V., OLIVE, D. I. & POLKINGHORNE, J. C. (1966b). *J. Math. Phys.* (to be published).
LANDSHOFF, P. V., POLKINGHORNE, J. C. & TAYLOR, J. C. (1960). *Proc. of the Rochester Conference on High Energy Physics*, p. 232.
LANDSHOFF, P. V., POLKINGHORNE, J. C. & TAYLOR, J. C. (1961). *Nuovo Cimento*, **19**, 939.

LANDSHOFF, P. V. & TREIMAN, S. B. (1961). *Nuovo Cimento*, **19**, 1249.

LANDSHOFF, P. V. & TREIMAN, S. B. (1962). *Phys. Rev.* **127**, 649.

LEE, B. W. & SAWYER, R. F. (1962). *Phys. Rev.* **127**, 2266.

LEHMANN, H. (1958). *Nuovo Cimento*, **10**, 579.

LOGUNOV, A., TODOROV, I. & CHERNIKOV, N. (1962). *Soviet Physics—JETP*, **15**, 891.

LU, E. Y. C. & OLIVE, D. I. (1966). *Nuovo Cimento* (to be published).

MACDOWELL, S. W. (1959). *Phys. Rev.* **116**, 774.

MANDELSTAM, S. (1958). *Phys. Rev.* **112**, 1344.

MANDELSTAM, S. (1962). *Ann. Phys.* **19**, 254.

MANDELSTAM, S. (1963a). *Nuovo Cimento*, **30**, 1127.

MANDELSTAM, S. (1963b). *Nuovo Cimento*, **30**, 1148.

MANDELSTAM, S. (1965). *Phys. Rev.* **137** B, 949.

MARTIN, A. (1964). *Strong Interactions and High Energy Physics* (Oliver and Boyd, ed. Moorhouse, R. G.).

MARTIN, A. (1965). CERN preprint 65/1857/5—TH. 637.

MATHEWS, J. (1959). *Phys. Rev.* **113**, 381.

MENKE, M. M. (1964). *Nuovo Cimento*, **34**, 351.

MØLLER, C. (1945). *K. Danske. Vid. Selsk.* **23**, no. 1.

MØLLER, C. (1946). *K. Danske. Vid. Selsk.* **22**, no. 19.

NAKANISHI, N. (1959). *Progr. Theor. Phys.* **21**, 135.

NAMBU, Y. (1957). *Nuovo Cimento*, **6**, 1064.

NAUENBERG, M. & PAIS, A. (1962). *Phys. Rev.* **126**, 260.

OEHME, R. (1961). *Phys. Rev.* **121**, 1840.

OEHME, R. (1962). *Nuovo Cimento*, **25**, 183.

OKUN, L. B. & RUDIK, A. P. (1960). *Nuclear Physics*, **15**, 261.

OLIVE, D. I. (1962). *Nuovo Cimento*, **26**, 73.

OLIVE, D. I. (1963a). *Nuovo Cimento*, **28**, 1318.

OLIVE, D. I. (1963b). *Nuovo Cimento*, **29**, 326.

OLIVE, D. I. (1964). *Phys. Rev.* **135** B, 745.

OLIVE, D. I. (1965). *Nuovo Cimento*, **37**, 1422.

OLIVE, D. I. & TAYLOR, J. C. (1962). *Nuovo Cimento*, **24**, 814.

OMNES, R. & FROISSART, M. (1963). *Mandelstam Theory and Regge Poles* (Benjamin, New York).

PERES, A. (1965). Unpublished.

POLKINGHORNE, J. C. (1962a). *Nuovo Cimento*, **23**, 360.

POLKINGHORNE, J. C. (1962b). *Nuovo Cimento*, **25**, 901.

POLKINGHORNE, J. C. (1962c). *Phys. Rev.* **128**, 2459.

POLKINGHORNE, J. C. (1962d). *Phys. Rev.* **128**, 2898.

POLKINGHORNE, J. C. (1963a). *Phys. Lett.* **4**, 24.

POLKINGHORNE, J. C. (1963b). *J. Math. Phys.* **4**, 503.

POLKINGHORNE, J. C. (1963c). *J. Math. Phys.* **4**, 1393.

POLKINGHORNE, J. C. (1963d). *J. Math. Phys.* **4**, 1396.

POLKINGHORNE, J. C. (1964a). *J. Math. Phys.* **5**, 431.

POLKINGHORNE, J. C. (1964b). *J. Math. Phys.* **5**, 1491.

POLKINGHORNE, J. C. (1965a). *Nuovo Cimento*, **36**, 857.

POLKINGHORNE, J. C. (1965b). *J. Math. Phys.* **6**, 1960.

POLKINGHORNE, J. C. & SCREATON, G. R. (1960 a). *Nuovo Cimento*, 15, 289.
POLKINGHORNE, J. C. & SCREATON, G. R. (1960 b). *Nuovo Cimento*, 15, 925.
REGGE, T. (1959). *Nuovo Cimento*, 14, 951.
REGGE, T. & BARUCCHI, G. (1964). *Nuovo Cimento*, 34, 106.
SCHWEBER, S. S. (1961). *Introduction to Relativistic Quantum Field Theory* (Harper and Row, New York).
SMITHIES, F. (1958). *Integral Equations* (Cambridge University Press).
SQUIRES, E. J. (1963). *Complex Angular Momentum and Particle Physics* (Benjamin, New York).
STAPP, H. P. (1962 a). *Phys. Rev.* 125, 2139.
STAPP, H. P. (1962 b). Berkeley preprint UCRL 10289.
STAPP, H. P. (1963). Berkeley preprint UCRL 10843.
STAPP, H. P. (1964 a). *Nuovo Cimento*, 32, 103.
STAPP, H. P. (1964 b). Berkeley preprint UCRL 11766.
STAPP, H. P. (1965). *High Energy Physics and Elementary Particles* (Intl. Atomic Energy Agency, Ed. A. Salam).
STREATER, R. F. & WIGHTMAN, A. S. (1964). *PCT, Spin, Statistics and All That* (Benjamin, New York).
SWIFT, A. R. (1965). *J. Math. Phys.* 6, 1472.
SYMANZIK, K. (1958). *Progr. Theor. Phys.* 20, 690.
TARSKI, J. (1960). *J. Math. Phys.* 1, 154.
TAYLOR, J. C. (1960). *Phys. Rev.* 117, 261.
TAYLOR, J. R. (1964). *Nuclear Physics*, 58, 580.
TAYLOR, J. R. (1966). *J. Math. Phys.* (to be published).
TER MARTIROSYAN, K. A. (1963). *Soviet Physics—JETP*, 17, 341.
TIKTOPOULOS, G. (1963 a). *Phys. Rev.* 131, 480.
TIKTOPOULOS, G. (1963 b). *Phys. Rev.* 131, 2373.
TRUEMAN, T. L. & YAO, T. (1963). *Phys. Rev.* 132, 2741.
UMEZAWA, H. (1956). *Quantum Field Theory* (North-Holland).
WANDERS, G. (1965). *Helv. Phys. Acta*, 38, 142.
WHITTAKER, E. T. & WATSON, G. N. (1940). *Course of Modern Analysis* (Cambridge) 4th edition.
WICHMANN, E. H. & CRICHTON, J. H. (1963). *Phys. Rev.* 132, 2788.
WIGHTMAN, A. S. (1960). *Dispersion Relations and Elementary Particles* (Wiley, Ed. De Witt, C. & Omnes, R.)
WU, A. C. T. (1961). *K. Danske. Vid. Selsk. Mat-fys. Medd.* 33, no. 3.
WU, T. T. (1961). *Phys. Rev.* 123, 678.
ZIMMERMANN, W. (1961). *Nuovo Cimento*, 21, 249.
ZWANZIGER, D. (1963). *Phys. Rev.* 131, 888.

INDEX

Where a topic is discussed on several consecutive pages, reference is made only to the first of those pages

Printed in the United States
By Bookmasters